M. OHRING

ELECTRON BEAM
ANALYSIS OF
MATERIALS

ELECTRON BEAM ANALYSIS OF MATERIALS

M. H. Loretto
Professor of Materials Science
University of Birmingham

LONDON
NEW YORK
Chapman and Hall

First published 1984 by
Chapman and Hall Ltd
11 New Fetter Lane, London EC4P 4EE
Published in the USA by
Chapman and Hall
733 Third Avenue, New York NY 10017

© *1984* M.H. Loretto

Printed in Great Britain at the University Press, Cambridge

ISBN 0 412 23390 8 (Cased)
ISBN 0 412 23400 9 (Science paperback)

This title is available in both hardbound and paperback editions. The paperback edition is sold subject to the condition that it shall not, by way of trade or otherwise, be lent, re-sold, hired out, or otherwise circulated without the publisher's prior consent in any form of binding or cover other than that in which it is published and without a similar condition including this condition being imposed on the subsequent purchaser.

All rights reserved. No part of this book may be reprinted, or reproduced or utilized in any form or by any electronic, mechanical or other means, now known or hereafter invented, including photocopying and recording, or in any information storage and retrieval system, without permission in writing from the Publisher.

British Library Cataloguing in Publication Data

Loretto, M.H.
  Electron beam analysis of materials.
  1. Materials – Testing 2. Electron beams
  – Industrial applications
  I. Title
  620.1'12   TA410

  ISBN 0–412–23390–8
  ISBN 0–412–23400–9 Pbk

Library of Congress Cataloging in Publication Data

Loretto, M.H.
  Electron beam analysis of materials.

   Bibliography: p.
   Includes index.
   1. Materials – Analysis. 2. Electron beams –
Industrial applications. 3. Electron microscopy.
I. Title.
TA417.23.L67   1984   620.1'127   84–11351
ISBN 0–412–23390–8
ISBN 0–412–23400–9 (pbk.)

CONTENTS

Preface	vii
Acknowledgements	viii

1 Introduction to electron beam instruments — 1
1.1 Introduction — 1
1.2 Basic properties of electron emitters — 1
1.3 Electron optics, electron lenses and deflection systems — 7
References — 18

2 Electron–specimen interactions — 19
2.1 Introduction — 19
2.2 Elastically scattered electrons — 19
2.3 Inelastically scattered electrons — 24
2.4 Generation of X-rays — 26
2.5 Generation of Auger electrons — 35
2.6 Generation of electron beam induced current and cathodoluminescence signals — 37
References — 38

3 Layout and operational modes of electron beam instruments — 39
3.1 Transmission electron microscopy — 39
3.2 Scanning electron microscopy — 44
3.3 Scanning transmission electron microscopy — 50
3.4 Auger electron spectroscopy — 53
3.5 Electron microprobe analysis — 56
3.6 X-ray spectrometers — 57
3.7 Electron spectrometers — 61
References — 64

4 Interpretation of diffraction information — 65
4.1 Introduction — 65
4.2 Analysis of electron diffraction patterns — 65

	4.3 Interpretation of diffraction maxima associated with phase transformations and magnetic samples	101
	4.4 Interpretation of diffraction patterns from twinned crystals	108
	4.5 Interpretation of channelling patterns and backscattered electron patterns in scanning electron microscopy	109
	References	112

5 Analysis of micrographs in TEM, STEM, HREM and SEM — 113

- 5.1 Introduction — 113
- 5.2 Theories of diffraction contrast in transmission electron microscopy — 114
- 5.3 Analysis of images in transmission electron microscopy — 119
- 5.4 Influence of electron optical conditions on images in TEM and STEM — 143
- 5.5 Interpretation of high resolution electron microscopy images — 144
- 5.6 Interpretation of scanning electron microscopy images — 147
- References — 151

6 Interpretation of analytical data — 153

- 6.1 Interpretation of X-ray data — 153
- 6.2 Interpretation of data from thin samples — 153
- 6.3 Interpretation of X-ray data from bulk samples — 162
- 6.4 Interpretation of electron energy loss spectra — 163
- 6.5 Interpretation of Auger spectra — 169
- 6.6 Spatial resolution of analysis — 175
- References — 177

Appendix A The reciprocal lattice — 179

Appendix B Interplanar distances and angles in crystals. Cell volumes. Diffraction group symmetries — 184

Appendix C Kikuchi maps, standard diffraction patterns and extinction distances — 189

Appendix D Stereomicroscopy and trace analysis — 198

Appendix E Tables of X-ray and EELS energies — 200

Index — 209

PREFACE

The examination of materials using electron beam techniques has developed continuously for over twenty years and there are now many different methods of extracting detailed structural and chemical information using electron beams. These techniques which include electron probe microanalysis, transmission electron microscopy, Auger spectroscopy and scanning electron microscopy have, until recently, developed more or less independently of each other. Thus dedicated instruments designed to optimize the performance for a specific application have been available and correspondingly most of the available textbooks tend to have covered the theory and practice of an individual technique. There appears to be no doubt that dedicated instruments taken together with the specialized textbooks will continue to be the appropriate approach for some problems. Nevertheless the underlying electron–specimen interactions are common to many techniques and in view of the fact that a range of hybrid instruments is now available it seems appropriate to provide a broad-based text for users of these electron beam facilities. The aim of the present book is therefore to provide, in a reasonably concise form, the material which will allow the practitioner of one or more of the individual techniques to appreciate and to make use of the type of information which can be obtained using other electron beam techniques.

The techniques which are covered in this book are electron diffraction, including convergent beam diffraction, conventional and high resolution transmission electron microscopy, scanning and scanning transmission electron microscopy, electron energy loss spectroscopy, Auger spectroscopy and X-ray microanalysis. In order to provide the necessary background information, the first chapter deals with electron optics at a very elementary level, the second chapter covers the factors which are important in influencing electron–specimen interactions and thus in generating the various signals which are detected. The third chapter covers the layout and mode of operation of the various instruments.

The theory has been kept to a minimum and references are given where appropriate to more detailed treatments. The book is aimed primarily at research scientists but it is written at a level appropriate to research students and final year undergraduates.

ACKNOWLEDGEMENTS

Many research students and colleagues have commented critically on earlier drafts of the book and I would like to take this opportunity of thanking them for their interest and help. I would also like to thank Miss Tina Salliss for her patience and skill in turning my difficult-to-read script into a typescript and Mrs E. Fellows for producing line diagrams and micrographs from originals of somewhat dubious quality. Finally, I would like to acknowledge the various authors and journals for allowing me to reproduce various figures in this book. These are acknowledged individually in the appropriate figure legends.

1
INTRODUCTION TO ELECTRON BEAM INSTRUMENTS

1.1 INTRODUCTION

Many electron beam instruments, which have been developed more or less independently of each other, are now being used either as individual or as hybrid instruments. Partly because the individual instruments have been developed separately, and partly because some instruments generate data which are more immediately interpretable, the fundamental relationships between the techniques and instruments tend to be obscured. The aims of this book are firstly to describe and discuss many of the modern instruments in a way which makes manifest the common ground between them and which highlights the advantages, disadvantages and fields of application of each instrument, so that their roles in materials science are clearly defined. Secondly, the book aims to provide the basic interpretation of the information which is available from each instrument, without attempting to present a detailed discussion of the underlying theory.

All of these instruments which are used to characterize materials have some similar basic requirements: an operating vacuum, although this varies between 10^{-3} and 10^{-10} Torr; an electron source; electron lenses for forming an electron probe; deflection systems for defining the probe position, and if necessary for rastering the probe; detectors to detect the signals, and an image-forming system. A discussion of vacuum systems is outside the scope of this book and in the introductory chapter, electron sources, electron lenses and deflection systems will be discussed; Chapter 2 deals with the basic science of electron–specimen interactions, the various instruments covered in the book are described in Chapter 3 together with the detection and processing of the many signals generated by electron–specimen interactions. The interpretation of signals is discussed in some detail in Chapters 4, 5 and 6. The appendices contain some essential data which are frequently required in interpreting the signals.

1.2 BASIC PROPERTIES OF ELECTRON EMITTERS

Electron sources in electron beam instruments are required to provide either a large total current in a beam of about 50 μm diameter, as in the case of low magnification transmission electron microscopy, or a high intensity probe of electrons as small as 0.5 nm in diameter, as in several scanning instruments. These requirements are not easily met by one type of electron source and sources have been developed which are more suitable for one type of application than for another. There are three basically different types of electron source available: the conventional tungsten hairpin filament (which can be modified to be a pointed tungsten filament), a lanthanum hexaboride crystal (LaB_6) and a field emission source.

In the presence of a suitable potential, electrons can be extracted from a source either by thermionic emission, in which the thermal energy of the electrons is sufficient for them to overcome the potential energy barrier (the work function) so that they can escape from the source, or by field emission which involves electron tunnelling so that the width of the potential barrier allows the quantum tunnelling of electrons so enabling them to escape from the source. Electron tunnelling requires very high field strengths and it is possible to operate sources in a hybrid manner so that the electron emission occurs at a lower field strength than is necessary for cold field emission and at a lower temperature than is necessary for thermionic emission, so that thermal field emission takes place. Before looking at the details of the different types of source it is useful to look at the characteristics of sources since it is these characteristics, the brightness, the stability, the size, the energy spread and the coherence, which define the performance of the various sources.

(a) Source brightness

The source brightness, β_s, is defined as the current density per unit solid angle and is measured in A cm^{-2} sr^{-1}. The brightness, which increases linearly with increase in accelerating voltage, determines the total current which can be focussed onto the specimen and thus determines the current in the small probes necessary in scanning instruments. Thus the maximum (theoretical) current, I_{th}, in a probe diameter d formed from a source of brightness β_s is given by [1]

$$I_{th} = \beta_s(\pi^2 \alpha^2 d^2)/4 \qquad (1.1)$$

where α is the semiangle of the probe-forming lens. Both the spherical aberration, caused by the probe-forming lens, and diffraction by the final aperture give rise to discs of confusion so limiting the current in a probe (see Section 1.3). The maximum current in a probe, subject to these two additional factors, is then given by

$$I_{max} = \frac{3\pi^2 \beta_s}{16}\left(\frac{d^{8/3}}{C_s^{2/3}} - \frac{4}{3}(1.22\lambda)^2\right) \qquad (1.2)$$

where C_s is the spherical aberration coefficient of the lens, λ the electron wavelength and d the probe diameter. For a fixed electron optical system

β_s is the only variable parameter and for the three types of electron sources, the conventional tungsten hairpin filament, the LaB_6 filament and the single crystal tungsten field emitter, the brightness varies roughly in the ratios 1, 10 to about 10^4.

(b) Source stability
Any variation of emission current with time is undesirable, but all electron sources suffer from short or long term instabilities to varying extents. Short term stability is important in scanned images, in order to avoid image flicker, and in scanned X-ray, Auger or other analytical applications where the level of the signal is significant. The tungsten hairpin filament is generally remarkably stable, with emission varying by less than $\pm 1\%$ over many hours, once any initial instabilities disappear. Typically these instabilities last for only a few minutes with a new filament and for an even shorter time when an old filament is switched on. It should be noted, however, that the steady emission current may change to a different steady current if the filament is switched off and on, and if a constant current is required over many hours, the filament should be left on even when specimens or film is changed. This requirement necessitates pre-pumped airlocks which are common in modern electron beam equipment. LaB_6 filaments are not quite as stable as the tungsten hairpin filaments but again, after the initial instabilities, the emission current is constant over tens of hours within $\pm 2\%$, unless the filament is overheated. Field emission sources are relatively unstable both over short time intervals – i.e. fractions of seconds – and over longer times, of the order of half an hour, and special operational modes are being increasingly introduced in order to compensate for these instabilities. These techniques are discussed in Section 1.2.1 where the layout of field emission sources is described.

The instability in the sources can be due to many causes: mechanical drift, ion bombardment of the filament by gas ions, adsorption of residual gases present in the microscope column, and whisker formation on the filament. The tungsten hairpin filament is not as sensitive to the vacuum level as are the LaB_6 and field emission sources; the hairpin filament requires a vacuum of better than about 10^{-3} Torr and it is nearly as stable at such a poor vacuum level as at a level of around 10^{-6} Torr which is required for a LaB_6 filament. The emission from a thermal field emission source requires a vacuum of the order of 10^{-7} Torr and a cold field emission source of about 10^{-10} Torr for stable operation, and in both cases the stability is better the better the vacuum.

(c) The source size
As discussed in Section 1.3 it is common practice to demagnify the electron source in order to reduce the significance both of lens aberrations and of non-homogeneous electron emission over the emitting area of the source.

In general it is accepted that a source should be used whose emitting area is greater than the largest area to be illuminated at any one time. Typical source sizes are roughly 50 μm for a conventional tungsten hairpin filament, 1 μm for a LaB_6 filament and about 5 nm for a field emission tip.

(d) Energy spread of sources
If there is a significant energy spread in the electron beam then chromatic aberration (see Section 1.3) associated with the lenses will lead to image degradation in transmission electron microscopy (TEM) and to an increase in probe size in scanning instruments. The influence of an energy spread, ΔE, is proportional to $\Delta E/E_0$ where E_0 is the accelerating voltage. The energy spread is caused both by instabilities in the HT source and by the inherent spread in energies associated with thermionic and field emission. The HT stability at 100 kV is typically 1 part in 10^6 leading to an energy spread of only 0.1 eV. On the other hand the energy spread associated with a thermionic source heated to about 2800 K is around 3 eV, the energy spread for a LaB_6 filament is about 1 eV, for a cold field emission source it is around 0.5 eV and for a thermal field emission source about 2 eV.

(e) Source coherence
The coherence of an electron source is a measure of the phase differences in the emitted beam and this controls the amount of interference which can take place between the direct and the various diffracted waves. For example the number of Fresnel fringes [3] formed in the image of a holey carbon film is very much higher for a microscope with a field emission filament than for a tungsten hairpin thermionic filament. This difference arises because the effective size of the source in a field emission tip is only about 5 nm so that the emitted electrons are in phase, i.e. the source is coherent, whereas the source size is about 50 μm for a tungsten hairpin filament.

The diameter of the specimen over which the illumination is coherent is given by λ/α where λ is the wavelength of the electrons and α is the semi-angle subtended at the specimen by the source. The coherence of a source is important for high resolution imaging (see Chapter 5) but is unimportant for all other applications discussed in the book.

1.2.1 Electron sources

(a) Conventional thermionic tungsten hairpin sources
A schematic diagram of a conventional tungsten hairpin source is shown in Fig. 1.1. The hairpin filament is heated to about 2800 K by direct resistance heating and the surrounding grid, known as the Wehnelt cylinder, together

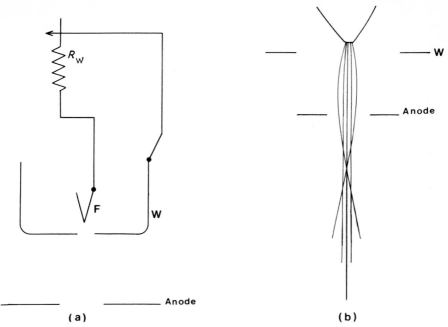

Fig. 1.1 Schematic diagram of a conventional tungsten thermionic source. (a) the filament F is at the accelerating potential of the instrument and is heated directly to about 2800 K. The Wehnelt cylinder W is biased by the potential drop across R_w (b) Schematic ray path showing focussing action.

with the anode, which is at earth potential, act as an electrostatic lens (see Section 1.3) forming an image of the region of the filament which is emitting electrons, just before or just beyond the anode aperture. This image of the filament is termed a crossover. The source size is about 50 μm, and because of the self-biassing action of the gun (see Fig. 1.1) the stability is excellent.

Measurement and calculation show that for typical operating conditions at 100 kV the brightness of such a thermionic gun is about 3×10^5 A cm^{-2} sr^{-1}. This type of gun, when used in a modern transmission electron microscope, is capable of producing an electron probe in the transmission electron microscope (TEM) mode (see Section 1.3) with a diameter d of 40 nm with a current of about 1 nA and in the scanning transmission electron microscope (STEM) mode (see Section 1.3) with $d = 4$ nm and a current of about 0.05 nA. Thus, although the conventional tungsten filament is capable of providing the relatively large total current ($\gtrsim 1$ μA) required for low magnification imaging in the TEM mode, it is not able to generate high intensity probes because of the inherent low brightness. Pointed single crystal thermionic tungsten filaments provide higher brightness sources than the hairpin filament but the recent success of LaB_6 filaments, which offer a greater increase in brightness and longer life, suggests that these filaments are a more satisfactory compromise between the thermionic and the field emission sources.

(b) LaB_6 sources
Although the development of LaB_6 filaments has been slow they are now replacing the conventional filament in many electron beam instruments. This has become possible partly because the vacuum in microscopes has been improved (mainly to reduce electron-beam-induced contamination of specimens caused by the high residual hydrocarbon level in the specimen area) so that LaB_6 filaments can be operated successfully, and partly because of the simplified design of the modern LaB_6 filament assembly. Basically the only difference between the conventional assembly illustrated in Fig. 1.1 and a modern LaB_6 assembly is that extra pumping holes are present in the Wehnelt cap to ensure a better pumping speed near the LaB_6 tip. The brightness of a LaB_6 filament can be as high as 10^7 A cm^{-2} sr^{-1} at 100 kV which is an improvement of a factor of about thirty over the tungsten hairpin filament. If the LaB_6 filment is operated at a lower brightness, say a factor five to ten above that of a tungsten hairpin filament, the lifetime can be increased to about 1000 hours, i.e. about ten times that of the tungsten filament. The higher current obtainable in small probes and the comparable total current from LaB_6 and conventional tungsten filaments, for use in low magnification transmission electron microscopy, will lead to the gradual replacement of the conventional tungsten filament by LaB_6.

(c) Field emission sources
A field emission source is usually a $\langle 111 \rangle$ orientation crystal of tungsten and the Wehnelt cylinder is raised to an extraction potential up to about 4 kV in order to cause emission from the tip of the crystal. The very high field around the tip focusses any gas ions present onto the tip and the consequent ion bombardment leads to short term instabilities in emission current. Thus the requirement for a high vacuum is clear but even at around 10^{-9} Torr the tip becomes contaminated so that the emission decreases and it is necessary to flash the tip periodically, i.e. to heat it up to drive off the adsorbed impurities.

In view of the short and long term instabilities in the emission, which influence any time-dependent signal, it is necessary to monitor the emission current (or a signal which is proportional to the emission) and compensate electronically for the fluctuations. This is usually done by using a signal derived from the (electrically isolated) aperture of the second condenser lens; for example the brightness of a scanning image can be scaled to compensate for changes in this current.

INTRODUCTION TO ELECTRON BEAM INSTRUMENTS

Table 1.1 Summary of the important properties of available electron sources.

Source	Brightness (A cm^{-2}sr^{-1})	Stability(%)	Source size	Energy spread (eV)
Tungsten	3×10^5	~1	50 μm	3
LaB$_6$	3×10^6	~2	1 μm	1
Cold field emission	10^9	~5	5 nm	0.5
Thermal field emission	10^9	~5	5 nm	2

The brightness of a cold or thermal emission source can be about 10^4 times that of a conventional tungsten filament. Because of the high brightness of field emission sources they are preferred in scanning instruments which require electron probes of 0.5 nm diameter but the small source size and associated small total current means that these sources are not ideal for low magnification transmission electron microscopy.

The important characteristics of the various electron sources are summarized in Table 1.1.

1.3 ELECTRON OPTICS, ELECTRON LENSES AND DEFLECTION SYSTEMS

1.3.1 Introduction

In this section a brief account will be given of the principles which underlie the operation of lenses in electron beam instruments and some important equations will be quoted so that the operation and limitations of probe-forming lenses, image-forming lenses and lenses used to collect and separate electrons of different energies may be understood. It is necessary to have at least this background in order to appreciate the operation of the various instruments discussed in Chapter 3 but it is outside the scope of this book to deal with the relevant electromagnetic theory; references are given at the end of this chapter to books which deal with this basic theory.

Electron lenses in electron microscopes are generally electromagnetic but the biased electron gun and lenses used in Auger spectrometers, and in some electron spectrometers attached to electron microscopes, are electrostatic lenses. The action and properties of electron guns are discussed in Section 1.2 of this chapter and a brief account of electrostatic lenses is given in Section 1.3.3.

1.3.2 Electron optics

The action of a magnetic field on an electron is described by the well-known right hand rule where the thumb, first and second fingers are used to represent the terms in a vector cross-product. The force **F** which an electron of charge $-e$ experiences when travelling with a velocity **v**, due to a magnetic field **B**, is given by

$$\mathbf{F} = -e(\mathbf{v} \wedge \mathbf{B}) \qquad (1.3)$$

and the magnitude of the force is then given by

$$F = Bev \sin \theta \qquad (1.4)$$

where θ is the angle between **B** and **v**. If the initial velocity of an electron is divided into two components, $\mathbf{v}_p$ parallel to **B** and $\mathbf{v}_o$ orthogonal to **B**, then the value of $\mathbf{v}_p$ will be unchanged by **B** (since θ will be zero) and the force resulting from **B** and $\mathbf{v}_o$ will result in circular motion of the electron about

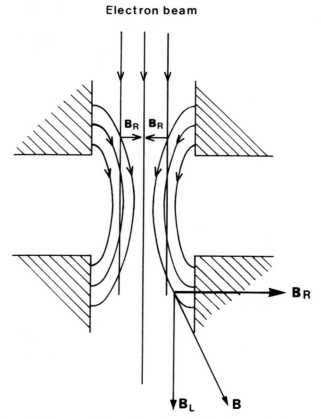

Fig. 1.2 Schematic diagram of the action of a cylindrical magnetic lens on the path of non-axial electrons. $\mathbf{B}_R$ is the radial component and $\mathbf{B}_L$ the longitudinal component of the field.

B. The radius of this motion is given by mv/eB and the resulting path of the electron will be a helix, the sum of the circular velocity and the unchanged longitudinal velocity, in the direction of the field.

If an inhomogeneous field is now considered we can see that it leads to a focussing action on electrons [2]. Thus, Fig. 1.2 shows, schematically, a cross section of a typical magnetic lens and the field produced by such a cylindrical lens is indicated in the figure. It can be considered to be made up of radial and longitudinal components, $\mathbf{B}_R$ and $\mathbf{B}_L$ respectively, which vary along the length of the lens, as shown schematically in Fig. 1.2. An electron entering the lens axially and precisely centrally will experience no force from the field in the lens, since the only component of the field in this part of the lens is $\mathbf{B}_L$ which is parallel to $\mathbf{v}$. For an electron travelling along the axis of the lens, but not passing centrally through the lens, the situation is very different. Such electrons will interact with the radial component of the field $\mathbf{B}_R$, since $\mathbf{B}_R$ is orthogonal to $\mathbf{v}$, and will experience a force of magnitude veB_R. This force causes the electron to change direction so that it now also experiences a force due to $\mathbf{B}_L$, the axial component of the field. The result of this is that the electron, which was initially travelling axially, spirals towards the centre of the lens, passes through the centre of the lens and then continues spiralling out from the centre before spiralling back. Such a trajectory is shown in Fig. 1.3.

It is easy to show [2] that the radial force on an electron is given by the expression

$$F = -\left(\frac{e^2}{4m}\right)B_L^2 r \qquad (1.5)$$

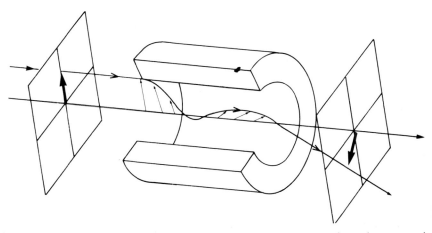

Fig. 1.3 Schematic diagram showing the trajectory of an electron through a magnetic lens.

where r is the radial distance of the axial electron from the centre of the lens. The force is proportional to r (hence the focussing action) and is directed towards the axis of the lens. Such lenses are always converging since F is independent of the sign of B_L.

The important parameters which describe electron trajectories within lenses, and which can be derived from the type of discussion presented above are: (1) the focal length of a lens and (2) the angular rotation of the image with respect to the object. If the axial extent of B_L is small, with respect to the focal length, then the (non-relativistic) thin lens formula for f, the focal length of the lens, can be shown [2] to be

$$\frac{1}{f} = \frac{300e}{8mc^2 E} \int_{\text{gap}} B_L^2 \, dL \qquad (1.6)$$

and the rotation of the image with respect to the object is given by

$$\theta = \left(\frac{300e}{8mc^2 E}\right)^{1/2} \int_{\text{gap}} B_L \, dL \qquad (1.7)$$

where E is the electron energy in electron volts, B_L is the longitudinal component of the field of the lens, m and e are the mass and charge (in esu) of the electron and c is the velocity of light. F is then given in centimetres and θ in radians. Since θ is controlled by B_L, successive lenses can be wound in opposite senses, such that θ_{total} is minimized or even reduced to zero.

The spherical aberration coefficient C_s of electromagnetic lenses is also calculable [2] and d_s, the diameter of the disc of confusion [3] referred to object space, is related to C_s by

$$d_s = 2C_s \alpha^3 \qquad (1.8)$$

where α is the semiangle aperture of the lens. Similarly the chromatic aberration coefficient C_c is related to d_c, the radius of the disc in object space, by the relation

$$d_c = 2C_c \alpha \frac{\Delta E}{E_0} \qquad (1.9)$$

where ΔE is the energy spread and E_0 is the accelerating voltage.

Typically C_s and C_c are of the same order as the focal length for a lens, and values for objective lenses in modern transmission electron microscopes are typically 1 to 2 mm and between 1 and 2 cm for scanning microscopes.

There is no convenient way with magnetic lenses to correct for spherical aberration and, unlike the case for optical microscopes, the only way to limit the influence of C_s on image quality and on probe diameter is to use an aperture to stop the lens down. Too small an aperture, however, degrades the information by diffraction from the aperture [3] and the diameter of the diffraction disc d_d is given by

$$d_d = 1.22\lambda/\alpha \qquad (1.10)$$

In probe-forming microscopes, where a suitable choice of the probe-forming aperture means that C_s is the limiting factor, it can be shown that there is an optimum aperture size for a probe of diameter d given by

$$\alpha_{opt} = \left(\frac{d}{C_s}\right)^{1/3} \qquad (1.11)$$

Similarly, by combining equations (1.8) and (1.10) linearly, the resolution which is given by the diameter of the disc of confusion is given by

$$d_{min} = A\lambda^{3/4} C_s^{1/4} \qquad (1.12)$$

The constant A is about unity. This equation can be used only as a rough guide for the resolution of an electron microscope since geometrical optics have been used. For $C_s = 2$ mm and 100 kV electrons this equation gives a resolution of about 3 Å.

1.3.3 Electron lenses

(a) Imaging and probe-forming lenses

The lenses in electron microscopes comprise a condenser system and an image-forming system and these lenses, together with the deflection coils (used for alignment and scanning) and apertures (which deliberately limit the collection or illumination angle) make up the electron optical systems which will be discussed here. Typical layouts of individual instruments are dealt with in Chapter 3.

The post-specimen lenses in TEM operate in a similar manner under all conditions of TEM because they are required to produce the total magnification simply by varying the power of one or more of the lenses. Of the imaging lenses it is the objective lens, which produces the first image of the object, which is required to be the most perfect lens. With modern 100 kV TEM/STEMs the values of C_s and C_c are typically about 2 mm, if 60° of specimen tilt is available, and 1.2 mm if only 15° is available. The value of C_s for high resolution electron microscopes (HREMs) operating at 600 kV–1 MV is about 2 mm even with 30° of tilt and because of the short wavelength the ultimate resolution is superior to that available in lower kV machines.

The condenser–objective lens systems in TEM/STEM instruments make most use of the flexibility of magnetic lenses, allowing a wide range of illuminating conditions to be used. The detailed modes of operation differ in different commerical instruments and the methods adopted in the Philips EM400T series will be used as a basis for the following description.

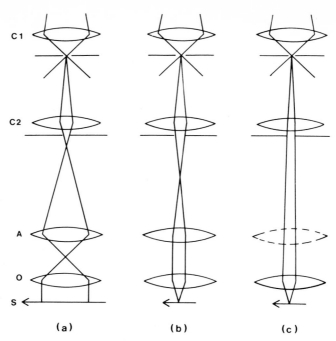

Fig. 1.4 Schematic ray diagrams in a transmission electron microscope. C1 and C2 are the first and second condenser lens, A is an auxiliary lens and O is the pre-field of the objective lens. S is the specimen. (a) illuminating system producing a near parallel beam at the specimen for TEM imaging; (b) production of convergent beam at specimen in TEM mode; (c) production of very small probe in TEM mode.

Fig. 1.4(a) shows schematically the illuminating system in a Philips EM400T operating in the TEM mode. In this mode the condenser–objective system provides a near-parallel beam at the specimen. This is achieved by operating the first condenser lens (C1) as a strong lens so that a demagnified image of the crossover from the electron gun is formed; the demagnification being of the order of a 100 × so that the 50 μm crossover, typical of a tungsten hairpin filament gun is reduced to less than 1 μm in diameter. The second condenser (C2), together with an additional auxiliary lens and the prefield of the objective lens, is used to project this image onto the specimen. In this type of microscope the specimen is immersed at the centre of the objective lens and the prefield plays an important part in the operation.

When C2 is overfocussed (Fig. 1.4(a)) a near-parallel beam illuminates a large area on the specimen. Typically the area can be as large as 25 μm in diameter, for low magnification observations. This can be reduced to only about 40 nm by suitably underfocussing C2, as shown in Fig. 1.4(b), where C2 and the auxiliary lens combine to produce a parallel beam, which is focussed onto the specimen by the objective prefield to a highly convergent probe. It should be noted that for conventional TEMs, which possess neither the auxiliary lens illustrated in Fig. 1.4 nor use the prefield of the objective lens as a significant part of the condenser system (because the specimen was not immersed in the objective lens), the smallest probe which could be produced at the specimen plane would be typically of the order of 2 μm in diameter. The ability to obtain a convergent probe as small as 40 nm in diameter is significant in microanalysis and convergent-beam or microdiffraction.

The convergence angle of the illumination (α_i) at the specimen, when C2 is overfocussed (Fig. 1.4(a)), is defined by the strength of C1 as can be seen by considering the smaller angular range of electrons from the source which would be accepted by the C2 aperture as C1 is strengthened (Fig. 1.4). To fill the C2 aperture completely it is evident that, as the strength of C1 is increased, the size of the C2 aperture must be decreased. As the strength of C1 is increased the intensity of illumination at the specimen, either for the parallel-beam mode or the probe mode, will be reduced since the illumination intensity is proportional to α_i^2 (see equation (1.1)). For a beam of minimum convergence, which is required for example for high angular resolution diffraction data, C1 should be made very strong which, with modern microscopes, can give a value of α_i of around 10^{-4} radians.

When the electron source is focussed by the condenser system the convergence angle is now controlled by the size of the C2 aperture, since a demagnified image of C2 is formed at the specimen surface. Increasing the strength of C1 decreases the probe current in this mode since, as pointed out above, fewer electrons now pass through the C2 aperture. The balance between C1 and C2, together with control of the size of the C2 aperture, leads to a very wide range of illumination conditions.

A third mode of operation of a TEM/STEM microscope is illustrated in Fig. 1.4(c). In this case the auxiliary lens is turned off and the strength of C2 reduced so that a very small probe is formed at the specimen by the objective prefield. The small probe size in this mode results from the fact that the crossover of C1 is imaged by the objective prefield, rather than the much closer crossover of C2. This mode of operation produces probes nearly as small as in the STEM mode.

In STEM a small probe is scanned over the area of specimen of interest (see Chapter 3). The small probe is obtained by increasing the strength of C1, changing the strength of the objective lens, switching off C2 and using an appropriately small C2 aperture to ensure that at the high excitation of C1 this aperture is filled with electrons. A probe as small as 2 nm can be produced at the specimen (see Fig. 1.5). For a given C2 aperture size the beam convergence is larger in the STEM mode than in the TEM mode since a more divergent probe is incident at the objective prefield/auxiliary lens. By weakly

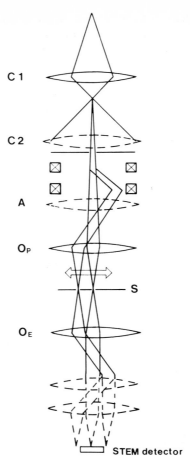

Fig. 1.5 Schematic ray diagram illustrating the mode of operation in STEM. C1 and C2 are the first and second condenser lens, A is the auxiliary lens, O_P and O_E the objective lens. The specimen is marked S. Post-objective lenses, used simply as transfer lenses in STEM are shown dashed.

exciting C2 and refocussing the probe using the objective lens the convergence angle in STEM can be continuously reduced with a corresponding increase in probe size.

In a scanning electron microscope (SEM) the probe-forming system is similar to that in STEM but in this case, because the probe-forming aperture defines the beam convergence and hence the depth of field of the microscope, the factors which control the selection of the size of this aperture are different from the STEM/TEM considerations discussed above. Generally the specimen is much further from the probe-forming lens in SEM than in STEM so that the focal length and hence the value of C_s for this lens is larger and the aberration limited probe size is therefore larger. The factors which are important in SEM in selecting the illuminating conditions are discussed in Chapter 3.

(b) Lenses in electron energy loss spectroscopy

A magnetic prism is used in TEM or STEM if it is required to separate electrons of different energies after the electrons have passed through the sample. As discussed in Chapters 4 and 5 the loss electrons can be used either to identify the elements in the specimen or to produce energy-filtered images or diffraction patterns. Typical ray paths are shown in Fig. 1.6 and the reason for the name magnetic prism is at once apparent. The action of the magnetic field on the electrons is to disperse them since the electrons which have lost the most energy are deflected through the largest angle. By varying the strength of the magnetic prism the whole desired energy range can be scanned over the slit and onto the detector so that a plot of energy loss versus intensity can be obtained. Alternatively the energy loss spectrometer

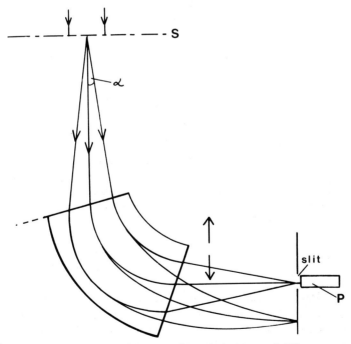

Fig. 1.6 Ray paths illustrating the separation of electrons of different energies in a TEM after they have passed through the specimen S and into a magnetic prism spectrometer where the electrons are ramped across the slit onto the photomultiplier P.

can be used as a STEM detector by presetting the strength of the magnetic prism and allowing only those selected electrons to produce the STEM image.

The parameter that defines the ability of the prism to separate the various energy electrons is the resolution. If the energy difference between two rays is ΔE and their spatial separation in the focal plane of the prism is Δ_k then the dispersion D is defined as

$$D = \frac{\Delta E}{\Delta_k} \qquad (1.13)$$

For a magnetic prism it can be shown [4] that

$$D = \frac{2R}{E_0} \qquad (1.14)$$

where R is the radius of the prism and E_0 the kV of the electrons. Thus if $E_0 = 100$ kV and $R = 50$ cm, $D = 10$ μm eV^{-1} the resolution would then be defined as $\delta = W/D$ where W is the slit width. However, magnetic spectrometers suffer from spherical aberration and the resolution is degraded accordingly, since the size of the disc of confusion is controlled by C_s. The resolution δ is given by [4]

$$\delta = E_0 \alpha \qquad (1.15)$$

where α is the solid angle accepted by the spectrometer (given by $\pi \alpha^2$, where α is indicated in Fig. 1.6) and E_0 the electron potential. Thus if α is 5 mrad and E_0 is 100 kV, $\delta \sim 7.5$ eV. Several different ways of improving the resolution can be used. Firstly α can be decreased, whilst maintaining, or even improving the collection angle at the specimen simply by appropriately coupling the spectrometer to the microscope (Chapter 3). Secondly the aberration of the spectrometer can be improved by shaping the pole pieces and thirdly the resolution can be improved by using a retarding field analyser (see next section) since the dispersion is inversely proportional to voltage. Using all or some of these techniques modern spectrometers have resolutions of about 1 eV even when the collection angle referred to the specimen is 50 mrad, the limit to the resolution being due to the energy spread in the electron source.

(c) Lenses in Auger spectroscopy

As discussed in Chapter 3 there are two different types of spectrometers used for collecting Auger electrons, one based on a cylindrical mirror analyser (CMA) and one based on a hemispherical analyser with a transfer lens (HSA). Both the transfer lens and the analyser are invariably electrostatic lenses rather than electromagnetic lenses and the voltage used is generally less than 1 kV.

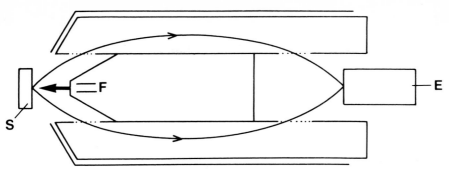

Fig. 1.7 Schematic ray diagram for the operation of a cylindrical mirror analyser (CMA). F is the filament, S the specimen and E an electron multiplier. The emitted Auger electrons are ramped across the multiplier by varying the potential across the analyser.

The action of electrostatic lenses is similar to that of electromagnetic lenses although the electrons are simply accelerated towards the region of relative positive potential without any rotation (cf. Fig. 1.3 for magnetic lens). In an analyser used in Auger spectroscopy the electrons enter the analyser at an angle to the lines of equipotential and are thus deflected from their original direction. A schematic ray diagram for a cylindrical mirror analyser is shown in Fig. 1.7. The potential across the cylinder is ramped so that the electrons are displaced across the extractor as a function of time and the signal which can be extracted from such an analyser is the intensity of electrons as a function of energy. The output is usually differentiated [5].

The transfer lens, used in conjunction with the HSA, is invariably a retarding lens which improves the resolution in the same way as can be done in electron energy loss spectroscopy (EELS). The retarding lens, which may consist of three electrodes, can be operated in two different modes. In the first mode the electrons from the sample are retarded to, say 1/20th of their initial energy and in this case the lens is operated as a two-electrode lens and the system will remain in focus as the spectrometer and lens are scanned to pass the different energy electrons onto the detector. In the second mode the energy of the electrons, which are allowed to pass through the analyser (see Section 3.7(b)), remains constant and, in order to achieve this, a potential has to be applied to the central electrode of the lens to focus the different energy electrons on the entrance slit to the analyser.

1.3.4 Deflection systems

Deflection systems are required in electron beam instruments both for alignment purposes and for scanning electron probes over the sample to produce scanning images. Beam shift and beam tilt controls are required

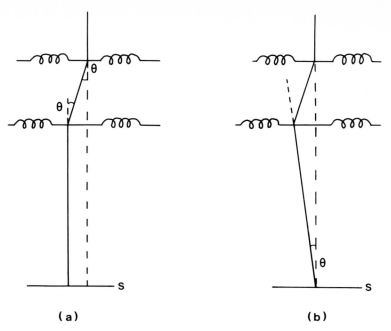

Fig. 1.8 Schematic diagrams illustrating action of: (a) beam shift; and (b) beam tilt coils. The specimen is at S.

and their different arrangements are illustrated schematically in Fig. 1.8. The field produced by the upper coils in Fig. 1.8(a) deflects the beam through an angle θ and the lower coils deflect the beam back through θ so that the beam is shifted without being tilted. In operations where the angle which the incident beam makes on the sample is critical, as in defect analysis in electron microscopy, it is clearly desirable to be able to align the beam without changing the angle of incidence. Fig. 1.8(b) shows how the tilt may be changed (as is required for example in dark field electron microscopy) without changing the part of the specimen which is being illuminated, i.e. without also shifting the illumination.

REFERENCES

1. Mulvey, T. (1967) in *Focussing of Charged Particles*, Vol. 1 (ed. A. Septier), Academic Press, London, p. 469.
2. Hall, C.E. (1953) *Introduction to Electron Microscopy*, McGraw-Hill, New York.
3. Jenkins, F.A. and White, H.E. (1951) *Fundamentals of Optics*, McGraw-Hill, New York.
4. Joy, D.C. (1979) in *Introduction to Analytical Electron Microscopy* (eds J.J. Hren, J.I. Goldstein and D.C. Joy), Plenum, New York.
5. Harris, L.A. (1968) *J. Appl. Phys.*, **39**, 1419.

2
ELECTRON–SPECIMEN INTERACTIONS

2.1 INTRODUCTION

The main requirement of all electron beam instruments is that they generate information which allows characterization of the sample. In order to interpret this information it is clearly essential that the origin of these signals, and the influence of all important experimental variables on them be understood. The generation of the signals which will be covered in some detail in this book, will be dealt with in turn in this chapter, together with the influence of important parameters, such as accelerating voltage and the atomic weight of the sample, on those signals.

2.2 ELASTICALLY SCATTERED ELECTRONS

Electrons which lose no, or virtually no, energy on interacting with the specimen are elastically scattered electrons. The elastic scattering of electrons by a crystal is best discussed by first considering the scattering by individual atoms, in terms of scattering by the nucleus and scattering by the electrons in the atom.

2.2.1 Elastic scattering of electrons by individual atoms

It can be shown (e.g. [1]) that the atomic scattering amplitude for electrons, $f(\theta)$, is given by

$$f(\theta) = \frac{m_0 e^2}{2h^2} \left(\frac{\lambda}{\sin \theta}\right)^2 (Z - f_X) \tag{2.1}$$

where m_0 is the rest mass of the electron and e the electron charge, h is Planck's constant, λ the electron wavelength, θ the scattering angle, Z the atomic number of the atom and f_X the atomic scattering factor for X-rays. For high accelerating voltages this expression must be corrected for relativistic effects by suitably scaling λ (see [1]) and by replacing m_0 by $m_0/(1 - v^2 c^{-2})^{1/2}$ where c is the velocity of light and v the electron velocity.

The part of the expression involving Z in equation (2.1) is the contribution

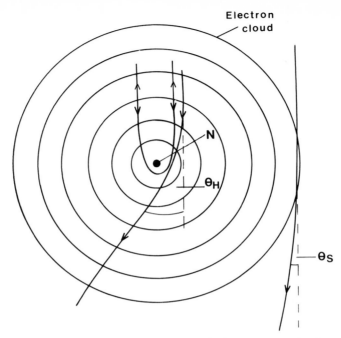

Fig. 2.1 Schematic diagram representing low and high angle scattering from an atom. The incident electrons follow the arrowed lines and the electron cloud contributes significantly to low angle scattering, θ_S, but the nucleus (N) is dominant at high angles (θ_H).

due to Rutherford scattering from the nucleus and the term involving f_X is the scattering contribution from the electron cloud. The term involving the electron cloud is more important at small scattering angles and the term involving the nucleus is dominant for large scattering angles. Fig. 2.1 is a schematic diagram which illustrates the difference between low and high angle electron scattering.

The atomic scattering amplitude $f(\theta)$ is shown plotted against $(\sin \theta)/\lambda$ for various values of Z in Fig. 2.2 and it can be seen that $f(\theta)$ decreases very rapidly with increase of angle. For example, the value for Au decreases by a factor of about four as $(\sin \theta)/\lambda$ increases from zero to about 0.5, which, at 100 kV ($\lambda = 0.037$ Å), corresponds to a scattering angle of about 1°.

Scattering factors for many elements have been calculated, using various approximations of the atomic electron energy distributions and the incident electron interaction with the innermost electrons. The most widely used are those given in [2]. The electron scattering amplitudes for heavy atoms are likely to be the least accurate because the approximations used are at their worst for heavy atoms.

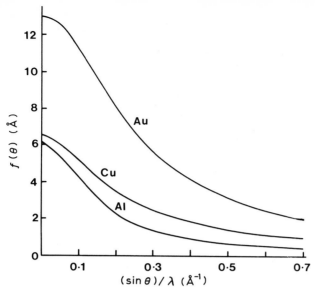

Fig. 2.2 Relationship between the atomic scattering factor $f(\theta)$ (in Å) and the scattering angle $(\sin\theta)/\lambda$ (in Å^{-1}) for aluminium, copper and gold calculated using equation (2.1).

The electron energy levels in crystals are modified, relative to those in isolated atoms, and these changes lead to changes in the atomic scattering amplitudes. It is clear that these changes will be more important for low angle scattering. In the case of ionically bonded solids scattering amplitudes can be calculated for the individual ions and (e.g. [2]) low angle scattering is modified when compared with the scattering from neutral atoms. For example at zero scattering angle the atomic scattering amplitude $f(\theta)$ is about 20% greater for Na$^+$ and Cl$^-$ ions than for neutral atoms and at $(\sin\theta)/\lambda \sim 0.1$ (0.2° at 100 kV) the atomic and ionic scattering factors are identical. In the case of metals the energy levels of the outer electrons are not accurately known and there is a corresponding uncertainty, which may be as large as 50%, for low angle scattering factors and about 10% for intermediate angles.

Scattering at high angles can be described adequately by considering only the term involving nuclear scattering and neglecting scattering by the electron cloud. The scattering angle at which scattering from the electron cloud can be neglected is given by [3]

$$\phi_R = 3.69 Z^{1/3} E_0^{-1/2} \tag{2.2}$$

where Z is the atomic number and E_0 the incident voltage. For 100 kV the values of ϕ_R for the elements Al, Cu and Au are 1.5°, 2.0° and 3.0° respectively.

These values should be borne in mind when assessing Fig. 2.1, remembering that at 100 kV the value of $(\sin \theta)/\lambda$ of 0.5 corresponds to only about $1°$.

2.2.2 Elastic scattering from crystals

(a) Low angle scattering

Scattering which gives rise to coherent diffracted waves is discussed here under the heading low angle scattering. Elastic scattering from crystals is of course influenced dramatically by the periodicity of the crystal and the smooth function of $f(\theta)$ versus scattering angle, shown in Fig. 2.1, is now inappropriate since strong diffracted rays are generated in specific directions defined by Bragg's law which is usually written in the form

$$n\lambda = 2d \sin \theta \qquad (2.3)$$

where n is an integer, λ is the electron wavelength, d the interplanar spacing and θ the angle the electrons make with the planes. In the present text n will be omitted and planes of spacing d/n considered. This is particularly useful when considering the reciprocal lattice as discussed in Appendix A and Chapter 4.

The structure factor for a reflection hkl, i.e. the amplitude of the reflection from the planes (hkl) is given by

$$F_{hkl} = \sum_i f_i(\theta) \exp \{ -[2\pi i (hu_i + kv_i + lw_i)]\} \qquad (2.4)$$

where u_i, v_i, w_i are the fractional atomic coordinates and $f_i(\theta)$ the atomic scattering amplitude for the atom i. The structure factors for face-centred cubic (f.c.c.) and body-centered cubic (b.c.c.) crystals are derived in Appendix A.

Because electrons are scattered so strongly by matter ($\sim 10^4$ as strongly as are X-rays) and because scattered electrons can be re-scattered, the relative intensities of diffracted beams, based on equation (2.4), will be in error for all but the thinnest crystals. Methods of determining the crystal structure using electron diffraction are therefore very different from methods using X-ray diffraction. Nevertheless it is possible (with care) to determine the Bravais lattice type by the presence of systematic absences in electron diffraction patterns. This is discussed in Chapter 4 and in Appendix A.

As discussed in Chapter 4 a crystalline sample can be oriented so that only one strong diffracted beam is excited. If this condition is fulfilled it is possible to calculate the number of atom planes which cause the incident electron beam to be reduced to zero intensity. This distance is defined as $\xi_g/2$ where ξ_g, the extinction distance, is given by [1]

$$\xi_g = \frac{\pi V_c \cos \theta}{\lambda F_g} \qquad (2.5)$$

where V_c is the volume of the unit cell, θ is the Bragg angle, λ the electron wavelength and F_g the structure factor for the particular reflection.

The accuracy of extinction distances calculated using equation (2.5) is limited by three factors. The first factor is the accuracy to which atomic scattering factors can be calculated and, as pointed out earlier, the errors are more significant the smaller the scattering angle. Certainly the accuracy will be no better than $\pm 10\%$. Second, the value of ξ_g derived from equation (2.5) is based on the assumption that only one diffracted ray is excited; many beam calculations can be carried out to correct for other beams if they are present and this becomes particularly important at high voltages, where systematic reflections of the sort $n\{hkl\}$ (where n is any integer) are inevitably excited (see Chapter 4). Finally, and very importantly, the precise deviation from the Bragg condition has a very important influence on ξ_g^{eff}, the effective extinction distance, as discussed in Chapters 4 and 5.

Bragg diffraction becomes less important for larger Bragg angles because as θ, the Bragg angle, gets larger the corresponding interplanar spacing gets smaller (cf. equation (2.3)), and the electron beam does not see the crystal as planar arrays of atoms. Thus only very small displacements, caused either by thermal motion of the atoms, or by defects, are needed to make the very closely spaced planes (which generate high angle diffracted beams) very irregular so that they no longer scatter coherently. The high angle diffraction information is thus smeared out by this mechanism. Cooling specimens is well known to increase the visibility of relatively high angle ($\sim 10°$) information (e.g. [4]). The intensities of Bragg diffracted beams are decreased by thermal motion by the Debye–Waller factor, $\exp(-2M)$ where M is given by

$$M = 8\pi^2 u^2 \frac{\sin^2 \theta}{\lambda^2} \qquad (2.6)$$

where u^2 is the mean square of the displacement perpendicular to the Bragg planes. The decrease in intensity of the corresponding Bragg peaks is compensated for by an increase in the diffuse background intensity and in fact the electrons which contribute to the background are one class of inelastically scattered electron as discussed in Section 2.3.

Thus, for the present purposes the division between low angle and high angle elastic scattering is made so that electrons which are likely to be Bragg diffracted are discussed separately from those which are scattered through too large an angle to be Bragg diffracted. Since the division between low angle and high angle scattering in this sense comes at not less than a scattering angle of about $10°$ at 100 kV it is clear that only Rutherford scattering need be considered when discussing high angle scattering (see equation (2.2)).

(b) High angle scattering

As pointed out above the term high angle scattering is taken to mean scattering which, in the ideal case, is not influenced by the periodic nature of the sample. The scattered intensity depends only on the nature of the atoms in the crystal and not on their relative positions, so that equation (2.1) provides a good description of the relation between the scattered amplitude and scattering angle. In fact only the term involving Z, i.e. Rutherford scattering, need be included, as pointed out above.

The Rutherford scattering cross section for deflection through an angle greater than ϕ may be written [5]

$$\sigma(\phi) = 1.62 \times 10^{-14} \frac{Z^2}{E_0^2} \cot^2\left(\frac{\phi}{2}\right) \qquad (2.7)$$

and for a layer of thickness dt the probability $P(\phi)$ of scattering through an angle ϕ is given by

$$P(\phi) = 1.62 \times 10^{-14} \frac{N\rho}{A} \frac{Z^2}{E_0^2} \cot^2\left(\frac{\phi}{2}\right) dt \qquad (2.8)$$

where N is Avagadro's number, Z the atomic number, A the atomic weight, ρ the density of the sample and E_0 is the incident accelerating voltage.

High angle elastic scattering is important because for example it is responsible for the degradation of the spatial resolution of X-ray ([6], [7], [8] and [9]) and Auger analysis and it also influences the resolution available in backscattered electron patterns and channelling patterns.

In summary it appears that elastic scattering can be modelled reasonably well over the two ranges of angle that are important in electron beam instruments. Thus, in transmission electron microscopy the relevant scattering angle defined by the objective aperture is about 0.2° so both scattering terms in equation (2.1) are important. In X-ray microanalysis scattering through angles greater than about 5° is the more important and in this angular range Rutherford scattering is dominant.

2.3 INELASTICALLY SCATTERED ELECTRONS

Electrons which lose energy when they interact with matter are by definition inelastically scattered. There are four mechanisms by which electrons lose energy when they are inelastically scattered. These are: (1) thermal diffuse or phonon scattering; (2) excitation of the electron gas or plasmon scattering; (3) single electron scattering; and (4) direct radiation losses. These will be discussed in turn.

2.3.1 Thermal diffuse or phonon scattering

Because atoms possess thermal energy they oscillate about their mean atomic positions and the intensities of diffracted rays are reduced by $\exp(-2M)$ as discussed in Section 2.2. The intensity lost from diffracted rays appears as diffuse background intensity. The incident electrons can gain or lose energy of the order of kT (~ 0.025 eV at room temperature) so the energy loss is minimal but the scattering angle can be very large. Formally this inelastic scattering process is due to the creation or annihilation of phonons. The mean free path for phonon scattering is about 1 μm for C and 200 Å for Au – a rapidly changing function with Z.

2.3.2 Plasmon scattering

The valence electrons in matter (conduction electrons in a metal) interact with fast electrons through a Coulomb interaction and many of the valence electrons are displaced for a very short time from their normal equilibrium positions. The fast electron excites collective oscillations, called plasmons, in the electron gas. The energy of the plasmon is given by [10]

$$E_p = \left(\frac{ne^2}{m}\right)^{1/2} \qquad (2.9)$$

where n is the number of valence electrons per unit volume of the specimen, e and m are the charge and effective mass of the electron.

Typically E_p, the energy loss suffered by the fast electron is of the order of 15 eV and the scattering intensity per unit solid angle has an angular half-width [10] given by

$$\theta_E = \frac{E_p}{2E_0} \qquad (2.10)$$

where E_0 is the incident voltage. θ_E is therefore about 10^{-4} radians. The mean free path for plasmon excitation is about 500–1500 Å [11]. Because this path length is small, plasmon loss peaks figure prominently in electron energy loss and Auger spectroscopy (see Chapter 6).

2.3.3 Single electron scattering

Energy may be transferred to single electrons (rather than to the large number, $\sim 10^5$, involved in plasmon excitation) by the incident fast electrons. Lightly bound valence electrons may be ejected, and these electrons can be used to form secondary electron images in SEM and a very large number of electrons with energies up to around 50 eV are ejected when a high energy electron beam strikes a solid. If inner shell electrons are ejected, i.e. if an atom is ionized, the appropriate energy must be lost by the incident electron. This characteristic energy lost by the incident electron is used in electron energy loss spectroscopy (EELS) to identify the atomic species. The probability of ionization and the actual energy lost by the incident electron are discussed in Sections 2.4 and 2.5 in terms of X-ray production and Auger electron production. Typically the mean free path for inner shell ionization is several micrometres and the energy loss can be several keV. It should be noted that the mean free path for ionization by ejection of inner shell electrons can be influenced by crystallinity and all the calculations of cross sections refer effectively to amorphous materials. Thus, as noted in subsequent chapters, there is increased electron absorption and X-ray generation when a crystal is oriented at an angle just less than the Bragg angle. In principle this effect can increase the probability of single electron scattering by a factor of three for very thin crystals [12] although in practice the observed increase is usually only about 20% because of the high incident beam convergence angle. It is nevertheless important to be aware of this effect during microanalysis and it is interesting to note that both elastic and inelastic scattering are influenced by the regularity of crystals. In addition to this complication it should be further noted that in ordered alloys the ratio of the intensities from the elements in the alloy can be changed dramatically if appropriate low order planes, rich in one component, are oriented just below the Bragg angle [13].

The angular half-width of scattering is given by $\Delta E/2E_0$ where ΔE is the energy lost and E_0 the accelerating voltage. Since the energy loss can vary from only tens of eV to tens of keV the angle can vary upwards from 10^{-4} radians.

2.3.4 Direct radiation losses

Fast electrons emit energy directly if they are decelerated and this energy forms the background radiation of the X-ray spectrum – bremsstrahlung. In principle an incident electron can lose all its energy in one event and this represents the highest possible energy of the X-ray background. The possibility of the generation of X-ray photons increases with decrease of photon energy as discussed in Section 2.4. The range of possible energy losses therefore ranges from zero to the incident energy and correspondingly the angular range of scatter is very large.

2.4 GENERATION OF X-RAYS

When electrons interact with matter X-rays will be generated. If the energy of the incident electrons is high enough to eject inner shell electrons then characteristic energy loss electrons will be generated and characteristic X-rays and Auger electrons will be emitted from the ionized atom as an outer shell electron falls into the inner shell vacancy. Thus a variety of characteristic energy X-rays is generated as the various displaced inner shell electrons are replaced by the various outer shell electrons. Clearly, the larger the number of electron energy levels which are present in the atom the larger is the number of different characteristic energy X-rays that can be generated. The characteristic X-rays are superimposed on a continuous background of X-rays (bremsstrahlung) which is generated when electrons are slowed down by interacting with the atoms in the target material. Because the origins of these two types of X-rays are so different the more detailed discussion of characteristic and continuous X-ray generation is dealt with in two sections.

2.4.1 Characteristic X-rays

The mechanism of generation of characteristic X-rays can be understood by reference to Fig. 2.3 where the various electron shells surrounding the nucleus are shown schematically. The shells are designated K, L, M ... corresponding to the principal quantum numbers 1, 2, 3 .... If a K electron is knocked out by an incoming electron the vacancy can be filled by an electron from either the L or M shell. If the electron falls in from the M shell a K_β X-ray is emitted which has an energy equal to the difference in energy between the two states of the atom. Similarly the generation of a lower

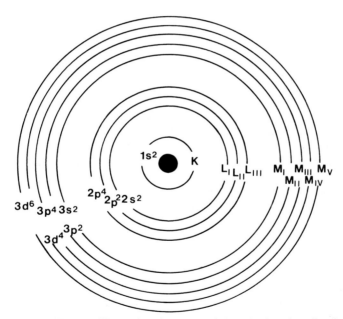

Fig. 2.3 Schematic diagram illustrating the nomenclature used to describe the electron shells using nickel as an example. The superscript gives the number of electrons in the K shell and the L, M, subshells, 1, 2 or 3 are the principal quantum numbers and s, p or d refer to the angular momentum quantum number.

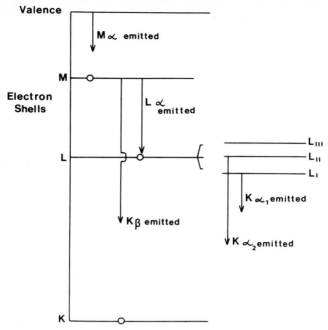

Fig. 2.4 Schematic diagram illustrating the various electron energy levels in nickel which give rise to K_α, K_β, L_α etc. X-ray lines when electrons fall between the shells indicated.

energy K_α electron takes place when an L shell electron falls into the K shell vacancy. The situation is summarized in Fig. 2.4. The relative probability of generating X-rays of the various possible energies from a given element is controlled initially by the incident electron energy and secondly by the number of different possible ways by which the ionized atom can return to its ground state.

This can be expressed more formally by the relationship,

$$n_E = Q\omega N I \quad (2.11)$$

where n_E is the number of X-rays generated of energy E, Q is the ionization cross section (i.e. the probability of ejecting the electron), ω is the fluorescent yield (i.e. the probability of the ionized atom returning to the ground state by emitting the specific X-ray photon), N is the number of atoms in the irradiated volume and I the incident electron flux.

The ionization cross section Q for any ionization is given by an equation of the form proposed by Bethe [14]

$$Q = \frac{\pi e^4}{E_0 E_c} Zb \ln\left(\frac{cE_0}{E_c}\right) \quad (2.12)$$

where E_0 is the incident accelerating voltage, E_c is the critical ionization energy, Z is the number of electrons in the shell and b and c are constants related to the details of the atomic structure. This equation is commonly written in the form

$$QE_c^2 = a(\ln cU)/U \qquad (2.13)$$

where U is the overvoltage ratio, E_0/E_c, and a and c are constants.

Mott and Massey [15] have calculated that $c = 2.42$ but Q then does not equal zero when $U = 1$ and there have been many attempts made to fit the observed voltage dependence of Q to experimental observations. The most widely used is that due to Green and Cosslett [16], where for $U < 10$ and $c = 1$ they find

$$QE_c^2 = 7.92 \times 10^{-14} (\ln U)/U \qquad (2.14)$$

where E_c is in eV, and Q is in cm^2. This equation is shown with QE_c^2 plotted against U for K shell ionization in Fig. 2.5. Other shells have similarly shaped curves but the constant is different [17].

At voltages used for STEM (~ 100 kV) a considerable extrapolation of these semi-empirical relations is required since U can be as large as 100. Constants for b and c in equation (2.12) have been determined [18] by matching data for Al, Ni and Ag with $b = 0.35$ and $c = 0.8U/[1 - \exp(-\gamma)][1 - \exp(-\gamma)]$ with $\gamma = \frac{1}{2}E_c$ and $\gamma = 1250/E_c U^2$.

The agreement is good for U between 4 and 25 and the values are similar to those obtained from equation (2.12) [18].

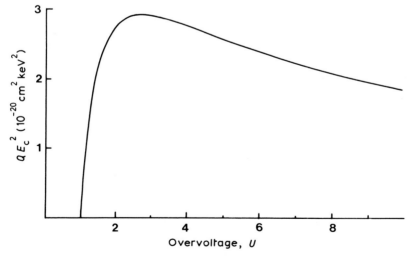

Fig. 2.5 Cross section for K-shell ionization calculated from equation (2.14) as a function of overvoltage, $U = E_0/E_c$.

From the present standpoint it is sufficient to note that the equations are empirical and that the values obtained for the cross section for K lines give reasonable agreement with experiment. The situation with L lines and other lines from outer shell ionizations is not as satisfactory and the significance of this in X-ray microprobe analysis is discussed in Chapter 6.

The variation of Q with voltage is particularly important in the analysis of bulk samples in the SEM or microprobe because those electrons, which are not backscattered out of the sample, lose energy by a combination of the various inelastic processes discussed earlier until their energy approaches that of the valence electrons when further scattering is irrelevant. Two other aspects of electron–solid interactions therefore require consideration if the generation of X-rays from bulk specimens is to be modelled successfully: the backscattering of electrons out of the sample and the change in Q as the electron loses energy by inelastic scattering processes.

Backscattering, i.e. large angle scattering, is of course high angle elastic scattering and has a higher probability in high Z material (equation (2.7)). The electron backscattering coefficient η is defined as the fraction of electrons that are scattered out of the sample. This can be determined fairly straightforwardly, experimentally, simply by measuring the fraction of incident current that flows to earth, and is very difficult to calculate. Typical values of η are shown in Fig. 2.6 which shows that η for bulk samples is a very sensitive function of Z.

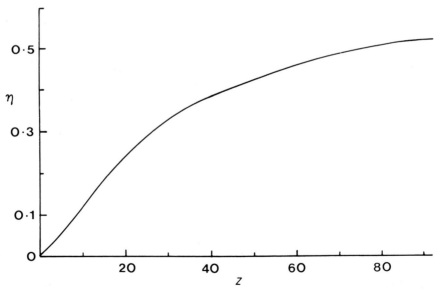

Fig. 2.6 Relationship between the backscattering coefficient η and atomic number Z for accelerating voltage of 10–30 keV [19].

The fractional loss in X-ray intensity η_x caused by backscattering requires a knowledge of the depth dependence of backscattering, the change in incident energy for each backscattered electron, as well as the exit path of the electron and any change in energy suffered on the way out of the sample; clearly an application of Monte Carlo modelling or other modelling techniques. Calculations and measurements have shown that Z dependence of η_x can be represented satisfactorily on families of curves of the type shown in Fig. 2.7 where $u^{-1} = E_c/E_0$ and η_x is plotted as a fraction of η. The factor by which backscattering reduces the X-ray intensity is commonly expressed as $R = (1 - \eta_x)$. It should be noted that for non-normal incidence η increases. For further details the reader is referred to specialist texts on microanalysis of bulk materials (e.g. [5]).

The change in Q as electrons penetrate a sample is best discussed in terms of the stopping power S of the sample, where S is defined as the rate of change of electron energy with distance traversed through the sample. Thus the number of ionization events dn in a path length dx is given for unit current by

$$dn = \frac{QN\rho}{A} dx \qquad (2.15)$$

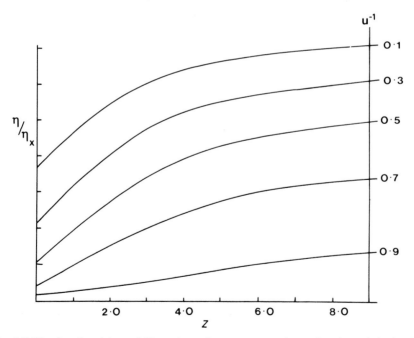

Fig. 2.7 The fractional loss of X-ray intensity η_x expressed as a fraction of the backscattering coefficient η plotted against atomic number Z for various values of U^{-1} where $U = E_0/E_c$.

where N is Avagadro's number, ρ the density and A the atomic weight of the sample. If it is assumed that electrons lose energy continuously as they pass through a solid, i.e. that the mean free path for inelastic scattering events is very small in terms of the total path length of the electron, it is possible to integrate equation (2.15) to calculate the total number of ionization events along the electron trajectory so that

$$n = \int_{E_c}^{E_0} \frac{QN\rho}{A} \frac{dx}{dE} dE$$

$$= \frac{N}{A} \int_{E_c}^{E_0} \frac{Q}{S} dE \qquad (2.17)$$

where S, the stopping power, is defined as $dE/d(\rho x)$. The Thomson–Whiddington law* can be used to calculate the rate of change of energy of an electron as it passes through a material of density ρ, thus

$$E_0^2 - E^2 = c\rho x \qquad (2.18)$$

in which $c = 3 \times 10^{11}$ eV^2 cm^2 g^{-1}.

By using this law it has been shown [16] that

$$n = 9.54 \times 10^{10} \frac{R}{Ac} (U_0 \ln U_0 - U_0 + 1) \qquad (2.19)$$

where R is the factor discussed above which allows for the reduction in X-ray yield caused by backscattering, and U_0 is the initial overvoltage, E_0/E_c.

On the basis of these empirical laws it is possible to obtain an assessment of the number of ionization events, although it must be emphasized that values of the constants R and S, which compensate for backscattering and stopping power, are composition dependent and appropriate weighting of values obtained for the constituent pure metals in alloys, must be used.

In order to obtain the total number of X-ray photons from equation (2.19) it is necessary to calculate the probability of an ionized atom returning to ground state by emitting an appropriate X-ray photon. The probability of an ionized atom returning to the ground state, by emitting a specific energy X-ray photon, is governed by two factors: the fluorescent yield and the relative probability of the various possible X-ray emissions.

The fluorescent yield ω defines the probability of an X-ray being emitted rather than an Auger electron (see Section 2.5). This yield is defined for the K shell by the equation

$$\omega_K = \frac{X_K}{X_K + A_K} \qquad (2.20)$$

---

* This empirical law relates the range of electrons to the voltage (cf. [20]).

where X_K and A_K are the number of X-ray photons emitted and the number of Auger electrons emitted. Calculations [20] show that the radiative transition probability is independent of atomic number Z whereas the probability of X-ray emission is approximately proportional to Z^4. Thus

$$\omega_K = \frac{Z^4}{a + Z^4} \tag{2.21}$$

where $a = 1.12 \times 10^6$ [3]. Alternatively the expression

$$\left(\frac{\omega_K}{1 - \omega_K}\right)^{1/4} = A + BZ + CZ^3 \tag{2.22}$$

has been used where values of A, B and C are fitted to experimental measurements. A curve of best fit for experimental measurements is shown in Fig. 2.8.

The value of the constant a in equation (2.21) for L_I lines is found to be 1.02×10^8 so that the X-ray yield is somewhat lower than for K lines and even lower yields are found for L_{II} and L_{III} lines.

The relative probability of the various possible electron transitions involved in X-ray emissions controls the relative intensities of the X-ray lines from an ionized atom. Considering, for example, the relative intensities of the K_{α_1} and K_{α_2} lines the electron transitions involved (see Fig. 2.4) are $L_{III} \rightarrow K$ and $L_{II} \rightarrow K$ respectively and consideration of the number of electrons

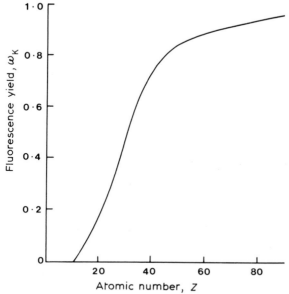

Fig. 2.8 Fluorescent yield ω_K for K lines plotted as a function of atomic number Z.

in these two subshells shows that there is twice the probability that an $L_{III} \to K$ transition will occur than an $L_{II} \to K$ transition. Hence if a K electron is removed by the incident electron the relative intensities of K_{α_1} and K_{α_2} will be 2:1 on this basis. A further complication is involved in assessing the relative intensities of X-ray lines. Thus the possibility of transitions within subshells (Coster–Kronig transitions) leads to redistribution of electrons within subshells, prior to radiation, and hence to a change in the relative intensities of the X-ray (and Auger) lines. This complication is discussed in several books on X-ray emission (e.g. [20]).

2.4.2 Bremsstrahlung

The continuous X-ray spectrum (bremsstrahlung) is generated together with characteristic X-rays when electrons interact with matter. The emission of the X-ray photons is associated with the slowing down of the incident electrons as they interact with the electrons and nucleii in the sample. The X-ray photons produced in this way will have all energies up to the energy of the incident electron, thus

$$eE_0 = h\nu_{max} = \frac{hc}{\lambda} \tag{2.23}$$

where e is the charge on the electron, E_0 the accelerating voltage, ν_{max} the maximum frequency, λ the corresponding wavelength of the X-ray photon, h Planck's constant and c the velocity of light. Insertion of numerical values shows that for λ (in Å) and E_0 (in eV)

$$\lambda_{min} = \frac{12.4 \times 10^3}{E_0} \tag{2.24}$$

Classical treatment shows [21] that the distribution of photon energies is given by

$$N(E) = \frac{aZ(E_0 - E)}{E} \tag{2.25}$$

where $N(E)$ is the intensity of X-rays of energy E, in photons per second per unit energy interval (in eV) per incident electron, $a \sim 2 \times 10^{-9}$, Z is the atomic number and E_0 the incident voltage. This equation shows that the intensity of the continuous background is proportional to Z and thus the characteristic X-rays, which are used in quantitative X-ray analysis, are superimposed on a larger background for specimens of high atomic number. From the present viewpoint bremsstrahlung is simply a background noise which must be removed from the characteristic signal. This is done in the main empirically, as discussed in Chapter 6, rather than by using an equation of the type shown in equation (2.25).

2.5 GENERATION OF AUGER ELECTRONS

If the energy of electrons which are incident on a specimen is high enough then, as discussed in Section 2.4, inner shell electrons will be ejected. The ionized atom, from which an inner shell electron has been ejected, may emit an Auger electron as it returns to the ground state, rather than emit an X-ray photon. The probability of an X-ray photon rather than an Auger electron being emitted is given by equation (2.21).

The Auger process is illustrated schematically in Fig. 2.9 where the particular Auger electron originates from a Si atom ionized by ejection of a K electron. The K vacancy is filled by an L_I electron and at the same time an L_{II}, L_{III} electron is emitted as an Auger electron. The nomenclature used to describe this specific process is KL_IL_{II}, L_{III}: the energy level from which the electron is ejected is named first, the level from which the electron falls named next, and the original level of the ejected Auger electron named last. After the emission of the Auger electron the atom is doubly ionized.

The probability of an ionized atom emitting a specific energy Auger electron is governed by two factors, the fluorescent yield (see equation (2.21))

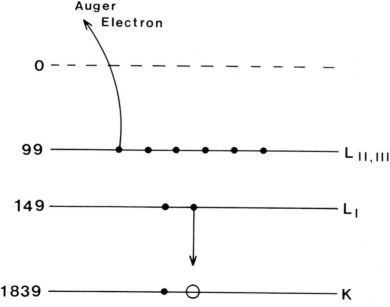

Fig. 2.9 Schematic diagram illustrating the emission of an Auger electron from an ionized silicon atom. The electron energy levels are listed on the left together with the corresponding X-ray nomenclature (see Fig. 2.3). The hole in the K shell in the ionized atom is filled by an L_I electron falling into this shell whilst an L_{II}, L_{III} electron is emitted as a KL_IL_{II}, L_{III} Auger electron.

and the number of different Auger electrons which can be emitted as an atom returns (eventually) to the ground state. To determine the number of transitions it is necessary to know the number of initial and final electron configurations where final refers to the doubly ionized state and not to the ground state. Consideration of the energy levels in doubly and singly ionized atoms [22] shows that for the KLL Auger transitions there are six Auger transitions for low and high Z atoms but nine for atoms of intermediate Z.

The energies of Auger electrons can be estimated by considering, for example, the process depicted in Fig. 2.9. The energy released when the L_I electron falls into the K shell is $(E_K - E_{L_I})$ but the electron has to supply the energy $(E'_{L_{II},L_{III}} + \phi)$, to escape from the sample, where ϕ is the work function. Because the atom has an extra positive charge, i.e. because it is doubly ionized, the term $E'_{L_{II},L_{III}}$ should be approximately equal to the L_{II}, L_{III} ionization energy of the next heavier element, so that $E'_{L_{II},L_{III}}(Z) = E_{L_{II},L_{III}}(Z + 1)$. Thus the energy of the Auger electron $E(Z)$ is given by

$$E(Z) = [E_K(Z) - E_{L_I}(Z)] - [E_{L_{II},L_{III}}(Z + 1) + \phi] \tag{2.26}$$

The measured energy will in fact have an additional term $-(\phi_A - \phi)$, which is the difference between the work functions of the energy analyser ϕ_A and the sample emitting Auger electrons. Thus

$$E(Z) = [E_K(Z) - E_{L_I}(Z)] - [E_{L_{II},L_{III}}(Z + 1) + \phi_A] \tag{2.27}$$

so that the work function ϕ is eliminated from the measurements.

Equation (2.26) is clearly limited since it predicts that $E_{KL_IL_{II,III}} \neq E_{KL_{II},L_{III}L_I}$ although the beginning and final states are identical so that the Auger electrons must have the same energies. Refinements of equation (2.26) taking into account quantum mechanical effects (e.g. [22]) remove this problem and give excellent (within 1%) agreement between experiment and theory.

The possibility of other transitions which can influence relative intensities of Auger and X-rays must be considered. The Coster–Kronig transitions are particularly significant because their transition rates are so high. These transitions are very rapid because they involve transitions within the one shell, e.g. of the type $L_{III} \rightarrow L_{II}$ and L_I and $L_{II} \rightarrow L_I$. The effect of this is to alter the relative intensities from those expected from the arguments given above, and in addition the energy available from these transitions may result in line broadening [23].

The influence of accelerating voltage on the generation of Auger electrons is felt through its influence on ionization cross sections discussed in Section 2.4. Thus the ionization cross sections, taken in conjunction with the fluorescent yields, and the probabilities of the various transitions, control the intensities of the Auger signals generated for a given beam current. The Auger signal generated by electron irradiation is superimposed on a high

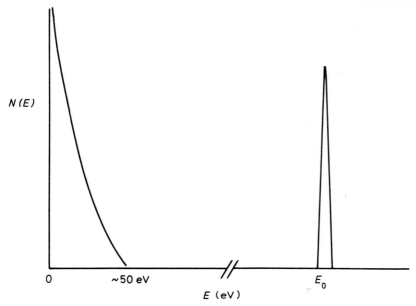

Fig. 2.10 Schematic diagram illustrating a typical $N(E)/E$ curve where $N(E)$ is the number of electrons of energy E emitted by a sample irradiated with incident electrons of energy E_0.

background signal made up of low energy secondary electrons (see Section 2.3) and a schematic output is shown in Fig. 2.10. It can be seen that there are basically three regions of electron energy in this figure: zero loss elastically scattered electrons; a region from zero energy to about 50 eV (the secondary electrons, see Section 2.3); and a region between these two. This region is where the Auger electrons and other loss electrons will be found, but they can be seen only if the vertical scale $N(E)$ is magnified, and/or differentiated.

2.6 GENERATION OF ELECTRON BEAM INDUCED CURRENT AND CATHODOLUMINESCENCE SIGNALS

These relatively novel signals are of particular interest in the study of semiconducting and insulating materials. Both signals require the production of electron–hole pairs. The electron beam induced current (EBIC) signal arises from the current produced by these electrons and holes as they diffuse, under an applied potential, though the sample (see Chapter 3). The presence of recombination centres reduces this current but when recombination occurs the resultant radiation may be able to be detected. If the band gap is appropriate, as in some crystals, the radiation will take the form of visible light which can of course be detected.

REFERENCES

1. Hirsch, P.B., Howie, A., Nicholson, R.B., Pashley, D.W. and Whelan, M.J. (1965) *Electron Microscopy of Thin Films*, Butterworth, Sevenoaks.
2. Doyle, P.A. and Turner, P.S. (1968) *Acta Cryst.*, **A24**, 390.
3. Wentzel, G. (1927) *Z. Phys.*, **40**, 590.
4. Steeds, J.W. (1979) in *Introduction to Analytical Electron Microscopy* (eds J.J. Hren, J.I. Goldstein and D.C. Joy), Plenum, New York, p. 387.
5. Reed, S.J.B. (1982) *Ultramicroscopy*, **7**, 405.
6. Kyser, D.F. (1979) in *Introduction to Analytical Electron Microscopy*, (eds J.J. Hren, J.I. Goldstein and D.C. Joy), Plenum, New York, p. 199.
7. Goldstein, J.I., Costley, J.L., Lorimer, G.W. and Reed, S.J.B. (1977) in *SEM/1977* (ed. O. Johari), IITRI, Chicago, p. 315.
8. Jones, I.P. and Loretto, M.H. (1981) *J. Microscopy*, **124**, 3.
9. Hutchings, R., Loretto, M.H., Jones, I.P. and Smallman, R.E. (1979), *Ultramicroscopy*, **3**, 401.
10. Egerton, R.F. (1976) *Phil. Mag.*, **34**, 49.
11. Raether, M. (1965) *Springer Tracts in Modern Physics*, Vol. 38, Springer, Berlin, p. 84.
12. Hall, C R. (1966) *Proc. R. Soc.*A, **295**, 140.
13. Bourdillon, A.J., Self, P.G. and Stobbs, W.M. (1981) *Phil. Mag.*, **44**, 1335.
14. Bethe, H.A. (1930) *Ann. Phys. Leipzig*, **5**, 325.
15. Mott, N.F. and Massey, H.S.W. (1949) *The Theory of Atomic Collisons*, Oxford, Clarendom Press.
16. Green, M. and Cosslett, V.E. (1961) *Proc. Phys. Soc.*, **78**, 1206.
17. Powell, C.J. (1976) *Rev. Mod. Phys.*, **48**, 33.
18. Zaluzec, N.J. (1979) in *Introduction to Analytical Electron Microscopy* (eds J.J. Hren, J.I. Goldstein and D.C. Joy), Plenum, New York, p. 121.
19. Bishop, H.E. (1968) *J. Phys. D:Appt. Phys.*, **1**, 673.
20. Dyson, N.A. (1973) *X-ray Analysis and Nuclear Physics*, Longman, Harlow.
21. Kramer, H.A. (1923) *Phil. Mag.*, **46**, 836.
22. Hörnfeld, O. (1962) *Ark. Fys.*, **23**, 235.
23. Chang, C.C. (1974) *Analytical Auger Electron Spectroscopy*, Plenum, New York, p. 509.

3

LAYOUT AND OPERATIONAL MODES OF ELECTRON BEAM INSTRUMENTS

3.1 TRANSMISSION ELECTRON MICROSCOPY

As its name implies the transmission electron microscope (TEM) is used to obtain information from samples which are thin enough to transmit electrons. The transmitted electrons are generally used to form either an image or a diffraction pattern of the specimen and schematic ray diagrams for these two modes of operation are shown in Fig. 3.1(a) and (b). The transmitted electrons can also be used for microanalysis since the characteristic energy loss suffered by electrons (see Chapters 2 and 6) can be used to identify the elements present. The use of a TEM and a STEM for electron energy loss is described in Section 3.7. When using the microscope in the conventional imaging mode, an objective aperture is inserted in the back focal plane of the objective lens. In the conventional imaging mode this aperture is used to select only one electron beam from which to form the image; a bright field image is formed if the directly transmitted beam is selected and a dark field image if a diffracted beam is selected.

In some cases the spacing of beams in the diffraction pattern makes it difficult to allow only the one beam through the objective aperture. For example, in the case of spinodal transformations (see Chapter 4) or the imaging of magnetic samples (see Chapters 4 and 5), closely spaced beams are generated and unless the aperture is deliberately positioned so that only one beam is allowed though the aperture, contrast may be observed due to interference of several beams in the image plane. It should be noted that the imaging of magnetic domains requires (see Chapter 5) that the specimen be positioned such that the sample is not saturated. This is achieved commonly by using a specimen holder so that the sample is held out of the strongest part of the field of the objective lens [1].

For high resolution electron microscopy (HREM) many diffracted beams, together with elastically and inelastically scattered electrons, which may appear between the diffracted beams in the diffraction pattern, must be allowed through the objective aperture if the periodicity of the sample and any departures from periodicity are to be resolved faithfully. This aspect is discussed in Chapter 5.

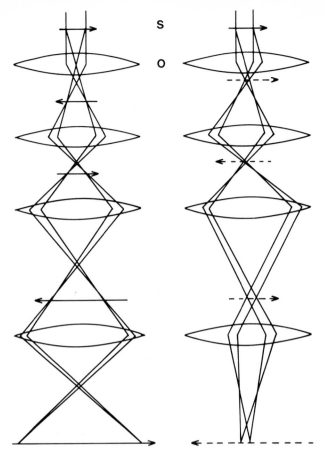

Fig. 3.1 Schematic ray diagrams for a transmission electron microscope with the specimen, S, illuminated with a parallel beam of electrons showing (a) image formation and (b) diffraction pattern formation by the objective lens, O, and the subsequent imaging lenses.

The objective aperture is removed to observe or record the diffraction pattern and the diffraction lens is focussed onto the back focal plane of the objective lens, rather than onto the first image plane. To define the area of the specimen from which the diffraction pattern originates a selected area aperture is inserted into the first image plane, but the area of the specimen defined by this aperture will not generally correspond precisely to the area from which the associated diffraction pattern originates. There are two sources of error. Firstly, if the plane of the selected area aperture does not coincide precisely with the first image plane then diffracted rays, originating from outside the area of the sample defined by the aperture, will contribute to the pattern. The extent of this selection error is given by $d\alpha$ where d is the focussing error and α the scattering angle of the electrons. Thus electrons scattered through large angles are more likely to originate from outside the area defined by the aperture. Since the aperture can be inserted in only the one fixed plane it is necessary that the first image is formed as precisely as possible at that plane to minimze d. Most modern microscopes have selected area diffraction (SAD) ranges of magnification where the first image is formed as accurately as possible at the plane of the selector aperture so that d is minimized.

The second source of error is caused by the spherical aberration of the objective lens which will cause off-axis rays to be brought to focus closer to the objective lens than are near-axial rays. Thus even if there is no focussing error, diffracted beams originating from areas outside that area defined by the aperture will contribute to the diffraction pattern. In this case a diffracted beam scattered through an angle α would be displaced at the first image plane by $MC_s\alpha^3$ where C_s is the spherical aberration coefficient of the objective lens and M the magnification of this image.

Depending on the sense of the focussing error the two factors giving rise to errors in SAD may or may not be additive. Since the magnitude of the term involving C_s depends on α^3 this tends to dominate SAD errors under normal operating conditions. For a scattering angle of 10^{-1} radians (which corresponds to a high order diffracted beam) and a C_s of 2 mm the error in specimen area is 2 μm.

The actual error in the area giving rise to the diffraction maxima observed in a SAD pattern is clearly a function of the scattering angle and an accuracy of about 1 μm would be typical for 100 kV microscopes. If an area smaller than this is selected by using a small SAD aperture it must not be assumed that all the diffracted beams originate from the area defined by the aperture. Clearly, by imaging successively in dark field with the diffracted beams the precise area giving rise to the individual beams can be defined. Despite this limitation the ability both to select an area of a specimen as small as 1 μm in diameter and to correlate image and diffraction information forms the basis of the success of TEM in the study of crystals – in particular in the study of crystal defects.

The provision of three or four post-objective lenses is now commonplace with conventional TEMs. This development has several important consequences: (1) magnifications of up to 10^6 can now be obtained on the final screen with sufficient intensity for focussing and correction of astigmatism; (2) the effective camera length (see Chapter 4) can be varied over a very wide range, whilst still retaining the selected area mode of operation; (3) the rotation between image and diffraction, caused by the necessary change in excitation of the diffraction lens, can be eliminated and rotation-free imaging obtained; (4) electron energy loss spectrometers (see Section 3.7) can be coupled to the microscope so that the efficiency of electron collection can be maximized; (5) in instruments which can be used in either the TEM or STEM mode the collection angle for STEM image formation can be varied to suit the particular requirements.

If rotation-free imaging is not designed into the microscope then the relationship between image and diffraction space can vary in a complex way as magnification and camera lengths are changed. It is therefore necessary to determine this relation and this is best done using a double exposure technique with a material such as MoO_3 where it is known that the long edges of the MoO_3 crystals are perpendicular to the smallest spacing in the diffraction pattern. The inversion, caused by the objective lens, between the first image and the diffraction pattern, may exist between the final image and final diffraction pattern. Whether or not this inversion is still present between the final image and final diffraction pattern is determined by the lens settings for image and diffraction. It is straightforward to find out experimentally whether a measured rotation of θ corresponds to $(\theta + 180°)$. This is done simply by underfocussing (or overfocussing) the diffraction lens so that the image of the specimen, which is visible in the out of focus diffraction pattern, can be compared directly with the conventional image. If the diffraction lens is underfocussed then the diffraction

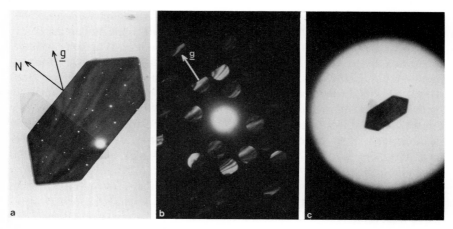

Fig. 3.2 A series of images taken to determine the rotation between image and diffraction space. (a) Double exposure of image and diffraction pattern with the rotation indicated. (b) Underfocussed diffraction pattern in which the extra rotation of the pattern caused by underfocussing can be seen. (c) Magnified central beam from (b) showing that this image of the specimen has been rotated by only a few degrees so that the rotation indicated in (a) is the true rotation for the specific lens settings used, i.e. there is no inversion.

pattern will not be inverted with respect to the normal diffraction pattern, and the presence of any inversion between this defocussed diffraction pattern and the normal image will be immediately apparent. This technique is illustrated in Fig. 3.2 where it can be seen that the total rotation between the underfocussed pattern (Fig. 3.2(b)) and the image is obvious. The slight rotation between the focussed and defocused diffraction patterns is also obvious (cf. Fig. 3.2(a) and (b)).

Most commerical microscopes operate between 100 and 200 kV but a significant number of high voltage electron microscopes (HVEM) (taken here to be greater than 200 kV) have been built, with 1 MV being the most popular choice. The layout of HVEMs is essentially identical to that illustrated in Fig. 3.1. A major difference between the 100 kV and 1 MV microscopes is in their physical size; HVEMs require much larger lenses and much more radiation shielding so that fork-lift trucks are required when changing lenses. The larger space available in HVEMs has led naturally to the development of stages which allow *in situ* experiments to be carried out, e.g. tensile stages and environmental stages. Nevertheless the standard specimens are still limited to 3 mm diameter discs, exactly as in 100 kV microscopes, and specimen holders are similar in HVEM and TEM. Tilting about two orthogonal axes of $\pm 60°$ and $\pm 45°$ is a common requirement in diffraction analysis but if the microscope is designed for HREM then the amount of tilt available is reduced. The tilting stages allow the selection of the many imaging conditions required for contrast analysis and for stereomicroscopy (see Appendix D), this latter technique being particularly valuable when examining thick specimens. Typically 1 MV microscopes allow the examination of samples which are about five times the thickness of those which can be examined at 100 kV. Lack of intensity, the increasing importance of chromatic aberration with increase of specimen thickness or a top–bottom effect [2] limit the penetration at 100 kV. Loss of intensity occurs because the various scattering mechanisms (see Chapter 2) combine to reduce the intensity of those electrons accepted by the objective aperture. If the size of the aperture is increased, in an attempt to increase the intensity, then more electrons which have lost larger amounts of energy will be accepted, since the scattering angle associated with inelastic scattering is of the order of $\Delta E/2E_0$, where ΔE is the energy loss and E_0 the accelerating voltage. Since the focussing error due to chromatic aberration is given by $\Delta f = \Delta E/E_0$, the increase in aperture size will lead to an increase in the chromatic error. At high voltages the value of $\Delta E/E_0$ decreases since E_0 is larger and, because the cross section for inelastic scattering decreases (Chapter 2), the fraction of electrons which suffer any given energy loss is smaller.

HREM leads to certain differences from conventional diffraction contrast microscopy. Thus the requirement that the gap between the pole pieces is small (since a small value of C_s is essential – see Chapter 1) leads to an increasing limitation on tilting facilities and to difficulties in having an energy dispersive X-ray (EDX) detector incorporated into the objective lens. There is therefore a tendency to regard HREMs as dedicated instruments which require high voltage and high coherency electron sources in order to reach the required resolution.

High resolution electron micrographs can be most easily interpreted if the sample is viewed (projected) precisely along an important crystallographic direction. This precision of alignment requires that regions as small as 50 nm in diameter be accurately tilted to specific beam directions. This is difficult using a selected area aperture (see earlier) and an objective lens capable of operating in a probe mode is perhaps going to be essential in future HREMs so that the orientation of such small regions can be accurately defined.

The most important difference between conventional electron microscopy and HREM is that in the latter high angle information must be allowed to interfere in the image plane in order to resolve the periodicity of the object, whereas in diffraction contrast TEM only one axial beam is allowed through the objective aperture. Thus an objective lens of very small spherical aberration is required in HREM so that the high angle information retains the correct phase relationship with respect to low angle information so that when they interfere in the image plane they reproduce faithfully the object periodicity. Additionally coherent illumination is required in order to obtain the correct phase relationships.

3.2 SCANNING ELECTRON MICROSCOPY

The SEM is used primarily for the examination of thick (i.e. electron-opaque) samples. Electrons which are emitted or backscattered from the specimen are collected to provide: (1) topological information (i.e. the detailed shape of the specimen surface) if the low energy secondary electrons ($\lesssim 50\,\text{eV}$) are collected; (2) atomic number or orientation information if the higher energy, backscattered electrons are used, or if the leakage current to earth is used. Imaging of magnetic samples using secondary and/or backscattered electrons reveals magnetic domain contrast. In addition two other signals can be collected, the electron beam induced current and light cathodoluminescence.

The output from any detector may be used to modulate the intensity of a cathode ray tube (CRT) which is scanned synchronously with the scan of the probe on the specimen. The magnification of the image is given immediately by the ratio of the CRT scan size to the specimen scan size. The image obtained in this way is simply the variation in intensity of the signal which is collected by the detector, as a function of position of the incident probe.

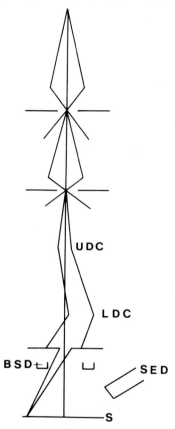

Fig. 3.3 Schematic ray diagram for a scanning electron microscope. The electron beam is rastered across the specimen, S. A backscattered detector (BSD) and secondary electron detector (SED) are indicated. The final probe-forming aperture is shown and the positions of the upper deflecting coils (UDC) and lower deflecting coils (LDC) are shown.

A schematic ray diagram for an SEM is shown in Fig. 3.3. The most commonly used electron detector is a scintillator–photomultiplier. If the low energy secondary electrons are to be efficiently collected a positive bias of 200 V is applied to the grid in front of the detector and if only the high energy backscattered electrons are to be collected the grid is biased to about -200 V, which will deflect virtually all the secondary electrons away from the detector. The high energy backscattered electrons will not be influenced by this small change in bias and the line of sight backscattered electrons will contribute to the image in both cases.

Solid state backscattered detectors can be fitted to the bottom of the probe-forming lens and are particularly useful in STEMs because they are so much smaller than scintillator–photomultiplier detectors.

If the electron probe is rocked through a range of angles about a fixed point on the specimen, then the angular dependence of the backscattered electron intensity can be obtained. If these electrons are used to modulate the intensity of the CRT then a channelling pattern is obtained (Chapter 4). Backscattered electron patterns can be obtained by setting the specimen so that the electrons strike the surface at a glancing angle. Electrons which are scattered through less than about 90° and subsequently Bragg diffracted, such that they re-emerge from the specimen (Chapter 4), then form a backscattered electron pattern. Because the probe is held stationary in this mode of operation the backscattered/Bragg diffracted electrons are detected using film, since no time-dependent signal is available. This technique thus requires a special camera to be inserted into the specimen chamber.

Because the signal is generally generated as a function of time in an SEM (in contrast to the TEM where the whole area of interest is illuminated simultaneously with electrons) it is straightforward to process images (e.g. to differentiate the signal, to back the signal off, etc.).

In addition to the secondary electrons or primary electrons it is possible to collect X-rays generated by the probe, and hence to determine local chemistry. Both wavelength and energy dispersive detectors can be used (see Section 3.6) and the analysis can be carried out either by placing the probe on the specific region of interest or by continuing scanning and using X-rays generated by the element of interest to modulate the intensity of the CRT. By selecting different energy X-rays for successive scans the distribution of specified elements can be obtained fairly rapidly. Unless the scan speed is slowed down significantly, these X-ray maps tend to be very noisy.

SEMs are also used in the cathodoluminescent mode and if the light emitted from the sample is used to modulate the CRT intensity, information on the distribution of the light-emitting regions can be obtained. This imaging mode is useful for studying light-emitting semiconductors.

In addition the beam-induced conductivity can be investigated by applying a potential across the specimen and using the current in this circuit to modulate the CRT (see Fig. 3.4). Clearly regions which give rise to significant increased conductivity will show different contrast from regions which give rise to little or no conductivity.

SEMs operate generally in the range 2.5 to 50 kV with probe sizes available at the specimen between 5 nm and 2 μm. The convergence angle of the probe at the specimen is controlled by the diameter of the final aperture (see Fig. 3.3) and this angle determines the depth of field of an SEM. Thus the large depth of field F which is commonly associated with SEM images is in fact due to the small convergence angle at the specimen, which is much

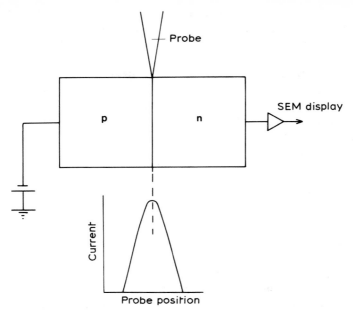

Fig. 3.4 Schematic diagram illustrating the use of an SEM to generate electron beam induced current (EBIC) information from a sample containing a p–n junction.

smaller than the corresponding angle in optical microscopes.* The relationship between the beam divergence α, resolution d (see below for the factors which are important in defining resolution) at which the SEM is being operated and F is easily seen to be given by $F = d/\alpha$ where F, the depth of field, is the distance between the point at which the probe is focussed and the highest/lowest point on the specimen at which the probe size exceeds d. If α is 5×10^{-3} radians and d is taken as 10 nm the value of F is 2 μm; this is a very large depth of field for such a high resolution image which underlines the value of high magnification SEM images of rough surfaces.

The resolution of an SEM is different for the various imaging modes and it is clear that at least three factors are important in determining the resolution: (1) the magnification and the number of lines on the CRT screen; (2) the probe size; and (3) the volume of specimen giving rise to the signal. These factors will be discussed in turn.

(1) The number of lines on the CRT is related to the maximum useful magnification by the expression $M = \delta/d$, where d referred to the speci-

---

* The requirement in optical microscopy that the collection angle of the objective lens be large, in order to obtain adequate resolution at high magnification, precludes the simultaneous satisfaction of the conditions for high magnification and large depth of field.

men is taken here as the probe size and δ is the screen resolution, defined by the number of lines. The number of lines is usually 1000 on a 10 cm screen so the lines are 0.1 mm apart (which is approximately the resolution of the human eye) so that $\delta = 0.1$ mm. The maximum useful magnification is given in this case by $M = 0.1/d$ if d is also measured in millimetres, and if M is 1000 the optimum probe size would be 100 nm. If a smaller probe size is used the image will be noisier than necessary and no gain in resolution will be achieved since the resolution is limited to 100 nm by the number of lines on the CRT at this magnification. Most manufactures provide charts which link magnification and probe size so that it is straightforward to choose the appropriate probe for the magnification required.
(2) The probe size is an important factor in determining the resolution of images obtained in an SEM and it seems reasonable that a probe of the same order as the required resolution should be used. However, especially at small probe sizes high angle elastic scattering (see Chapter 2) degrades the resolution significantly.
(3) As pointed out above, the volume of the specimen is important in determining resolution because electron scattering takes place within the sample. The extent of the degradation of the resolution is different for the different signals and these will therefore be considered separately.

The secondary electrons, which give rise to the topographical information,

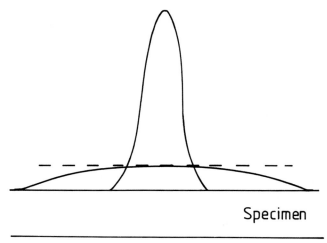

Fig. 3.5 Schematic diagram illustrating the intensity of secondary electrons generated from a solid specimen by a stationary probe. The dashed line represents the level to which the background must be set to remove any contribution to the secondary image from secondaries generated by backscattered electrons outside the initial probe diameter.

are generated both by the incident electrons as they enter the specimen and by backscattered electrons and X-rays. Fig. 3.5 shows schematically the form of the variation in intensity of the secondary electron signal as a function of distance away from the incident electron probe. The central peak in intensity is generated by the incident electrons, and therefore has a peak shape which reflects the variation in intensity across the incident probe. The full width at half maximum height (FWHM) of the incident probe is thus a measure of the FWHM of the main secondary electron signal. This signal is superimposed on the much weaker secondary electron signal generated by electrons which have been either singly elastically scattered, so that they travel significant distances close to the surface of the sample, or multiply scattered so that they are backscattered from the sample outside the area defined by the initial probe. The intensity of this secondary electron signal generated by high angle elastically scattered electrons is clearly going to be far smaller than that of the signal generated within the area of the probe because (apart from those few high energy electrons scattered such that they travel close to the surface), all backscattered electrons which give rise to secondary electrons, which are close enough to the surface to escape, must suffer at least two high angle elastic scattering events. The probability of an electron being scattered through an appropriate large angle by successive high angle scattering can be seen to be very small from Fig. 2.2 or from equation (2.8). The intensity of the secondary electron signal, generated by high angle elastically scattered electrons, will be reduced with respect to the main secondary signal, by the product of the probabilities of scattering through the two angles. Similarly the intensity of secondary electrons generated by X-rays (which are generated by the incident high energy electrons) will be small because the probability of X-ray generation is small (see Chapter 2) and the probability of subsequent secondary electron generation, near the surface of the sample, is also small.

On the basis of the above discussion the resolution of a secondary image in an SEM is potentially significantly degraded by high angle elastic scattering. The fact that secondary images can be obtained in which detail is visible on a scale of the FWHM of the incident probe is due to the ability to process images. Thus if the background level is adjusted so that the information visible on the CRT corresponds to intensities above the dashed line on Fig. 3.5 the secondary electrons generated by the backscattered electrons will not contribute to the secondary image and the resolution will be virtually that defined by the incident probe size. In principle the resolution can be improved even further by raising the background level to a higher level than indicated on Fig. 3.5 so that image detail which is smaller than the probe at FWHM can be observed. The ultimate limit to resolution is defined by signal-to-noise considerations since this sets the limit to the initial probe size and the limit to the extent to which the background level can be raised. An ideal sample consisting say of gold particles on aluminium would generate a large secondary signal from the gold and a small backscatter-generated secondary signal from the aluminium substrate and in such a sample resolution could be better than the initial probe size. With more typical samples the optimum resolution of secondary images would be about 5 nm but the above discussion should make it clear that resolution is a somewhat difficult parameter to define in this imaging mode.

The spatial resolution of backscattered electron images is also affected by electrons which suffer two or more high angle scattering events and which emerge from the specimen travelling in a direction such that they are collected by the backscattered electron detector. The intensity profile of backscattered electrons will therefore appear qualitatively similar to that for secondary electrons illustrated in Fig. 3.5. Similar arguments apply in terms of adjusting the background level and thus rejecting the electrons originating from well outside the area illuminated by the probe. However, the backscattered signal is not as intense as the secondary image and the spatial resolution in a medium atomic weight specimen would be about 25 nm. Again it is difficult to put a precise figure on the resolution because it is influenced significantly by the specimen.

Similarly the specimen current mode is limited both by beam spreading and relatively noisy electronic amplification to a spatial resolution of about 0.5 μm.

The spatial resolution of X-ray microanalysis is degraded more significantly than are secondary and backscattered images and a typical figure for X-ray resolution would be about 2 μm. In this case electrons have to be scattered through a high angle only once, after which they may degrade the X-ray resolution if they ionize an atom. Characteristic X-rays are generated with spherical symmetry and at whatever depth in the sample they are generated some will contribute to the signal which is detected, unless they are of very low energy in which case they may be absorbed. Electrons which have lost a significant fraction of their initial energy, and could not therefore be backscattered out of the sample, can nevertheless generate X-rays, so that the spatial resolution of X-ray signals will be expected, on this basis also, to be worse than secondary and backscattered images.

Somewhat similar arguments apply to cathodoluminescence and beam induced conductivity and these techniques have typical resolutions of about 2 μm. Indeed it is the limitation of the spatial resolution of SEM which has been responsible in part for the development of STEM which is discussed in Section 3.3.

3.3 SCANNING TRANSMISSION ELECTRON MICROSCOPY

As its name implies the STEM is both a scanning microscope and a transmission microscope. Essentially two different types of STEMs have been developed which operate* between 100 and 200 kV. The first type of these can be operated either in the TEM mode or in the STEM mode in which a scanned probe is used to generate image information, whereas the second type can be operated only in the STEM mode; this latter type of STEM is commonly referred to as a dedicated STEM, but if post-specimen lenses are added this removes much of the significant difference and the only remaining differences between the two types of STEMs are as follows. (1) The TEM/STEM microscopes can use film to record all image and diffraction information directly (other than the STEM image) whereas all image and diffraction information is obtained in a dedicated STEM by photographing either a CRT or a fluorescent screen. This is a significant limitation of dedicated STEMs and it is likely that improved image recording will become available. (2) The general vacuum technology is better in dedicated STEMs because they always use field emission guns which require a vacuum of about 10^{-9} Torr. Field emission guns are available for TEM/STEMs.

The electron beam is focussed to a small probe when operating in the STEM mode and this probe is scanned across the specimen. With a thin specimen, transmitted electrons can be collected by a transmission detector (cf. Fig. 1.5) and the intensity of these electrons used to modulate the intensity of a synchronously scanned CRT in a similar way to the method used in a SEM. The magnification is again obtained from the ratios of the scan sizes on the CRT and on the specimen. The lenses after the objective lens are thus not used to magnify, they simply transfer information to the detector, which is in a plane conjugate with the back focal plane of the objective lens. Any distortion of the probe shape caused by the post-specimen lenses is unimportant as long as the appropriate electrons are collected by the detector, since it is their integrated intensity which is used to modulate the CRT intensity.

If a bright field image is to be formed then the STEM detector must collect only the direct beam. If a dark field image is required then an annular detector can be used and any selected off-axis rays can be collected to form the image. The detector is of a fixed size and the collection angle of the detector can be varied simply by changing the effective camera length; at small camera lengths the bright field detector may collect diffracted beams as well as the direct beam but as the camera length is increased the magnification of the diffraction pattern increases and the detector will collect fewer electrons. The camera length can be made so large that the bright field detector will collect only a fraction of the direct beam; this can be a useful way of avoiding oversaturating the detector and of obtaining images which contain diffraction contrast information (see Chapter 5) even when a high probe current is being used to increase the X-ray count rate.

* Some 1 MV STEMs are being built but the high cost of these will inevitably limit the number which are used and more recently 300–400 kV instruments have become available commercially.

Post-specimen coils can be used to scan the diffraction pattern over the STEM detector and hence the pattern is displayed on the CRT screen. This can be done when operating in either the TEM or STEM mode and if the Y-modulation is used an immediate intensity map of a diffraction pattern is obtained. Alternatively the STEM detector can be removed (if it is above the viewing screen in a conventional microscope) and the diffraction pattern observed on the final screen. This diffraction pattern can then be photographed so that it is recorded directly on film. The area to which the diffraction pattern corresponds is clearly that area of the specimen being scanned, i.e. the area imaged on the STEM CRT. Clearly if the scan is stopped at a specific point on the STEM and the diffraction pattern imaged by removing the STEM detector then the diffraction pattern corresponds to that area of the specimen on which the probe is stopped since that is the only part of the specimen being illuminated with electrons. With probe sizes down to around 1–5 nm this technique offers a vast improvement in the accuracy of selected area diffraction over that available in TEM (cf. Section 3.1). The fact that STEM diffraction patterns can be photographed directly on film means that detail is not lost. This is particularly important when considering convergent beam diffraction patterns, which are discussed in Chapter 4.

As the name implies convergent beam diffraction patterns are obtained with a convergent rather than with an approximately parallel electron beam. The discussion in Chapter 1 shows that in order to maintain sufficient electrons in a probe the convergence angle must be increased as the probe size decreases so that convergent beam diffraction patterns are inevitably formed when operating in the STEM mode. With TEM/STEM instruments there are essentially two different modes of operation to obtain convergent beam diffraction patterns. In the first mode, the TEM mode, the microscope is set up in the imaging mode and when a focussed image has been obtained, the second condenser lens is used to focus the probe onto the specimen. A convergent beam diffraction pattern can then be viewed on the final screen by selecting the diffraction mode, and removing the objective aperture. The convergence angle is then defined by the size of the second condenser aperture since this mode of operation simply focusses this aperture onto the specimen; the probe size on the specimen and the current in the probe are controlled by the strength of the first condenser (see Fig. 1.4).

In the second mode of operation the probe size and convergence angle are controlled by both the second condenser lens and the objective lens. For the largest convergence angle the second condenser is switched off and the probe focussed using the first part of the objective lens. In this mode the second condenser aperture defines the beam convergence and the excitation of the first condenser lens controls the probe size; this mode of operation is commonly referred to as the STEM mode (see Fig. 1.5). If the excitation of the second condenser is now increased and the excitation of the objective lens decreased the beam convergence at the specimen will decrease and the probe size will increase. It is thus possible to obtain a complete range of beam convergence angles even with the usual limited range of condenser apertures but the increase of probe size with decrease of convergence limits the value of this technique somewhat. As discussed in Chapter 4, the information in convergent beam diffraction patterns is very dependent upon thickness and crystal perfection and the use of a large probe results in relatively featureless patterns because thickness and perfection are then likely to vary over the volume sampled, and the information is averaged.

Since the information in STEM is obtained sequentially the image can be processed in exactly the same way as in SEM. X-ray microanalysis is also an integral part of virtually all STEM facilities and because electron-transparent specimens are used, the degradation of the spatial resolution of X-ray signals is very small compared with that in the bulk specimens used in SEM. This aspect is discussed in Chapters 2 and 6.

Because of the low X-ray fluorescent yield for light elements and because of the increased absorption which takes place both within the specimen and in the beryllium window, EDX is not useful for elements below sodium in the periodic table (see Chapter 2). The microanalytical facilities in a STEM are greatly improved by the addition of an electron energy loss spectrometer (EELS). Elements from lithium upwards can be detected using EELS and this technique is being increasingly used.*

Finally, in a STEM it is possible to collect secondary electrons and/or backscattered electrons so that all the techniques available in a SEM are available in STEM – the major difference being of course the limitation on specimen size imposed by the limited volume available in the objective lens of a STEM.

3.4 AUGER ELECTRON SPECTROSCOPY

Auger electron spectroscopy (AES) is an important technique for studying the chemical composition of surfaces. As discussed in Chapter 6, typical escape depths for Auger electrons do not exceed 3 nm and are commonly below 1 nm. Measurement of the energies of Auger electrons thus provides surface chemical information, and this combined with successive removal of surface layers by ion beam milling (sputtering) enables composition depth profiles to be obtained. Other closely linked techniques, secondary ion mass spectrometry (SIMS) and electron spectroscopy for chemical analysis (ESCA), are commonly incorporated into multipurpose Auger facilities,

---

* It should be noted that windowless EDX detectors, or detectors with ultrathin windows, are being interfaced to STEMs and SEMs with the result that EDX analysis down to beryllium is possible despite the low fluorescent yield.

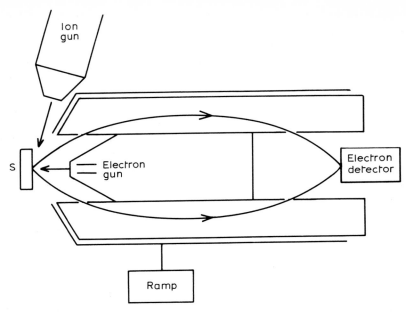

Fig. 3.6 Block diagram illustrating the layout of an Auger electron spectrometer showing the specimen S, the cylindrical mirror analyser (CMA), ion gun and electron detector. The potential across the CMA is ramped so that different energy Auger electrons are collected successively on the electron detector.

but these technique of surface analysis will not be discussed here.

A schematic layout for an Auger spectrometer is shown in Fig. 3.6. An electron gun operating at voltages up to 10 kV is used to bombard the specimen. The condenser system incorporated into the gun allows probe sizes, at the specimen, of between about 50 nm and 5 μm to be used. The electron probe can be scanned over the sample and a secondary electron detector (or leakage current) can be used to generate a scanning image as in a SEM. If the probe is stopped at a selected point then the Auger electrons generated from that region can be collected by the electron energy analyser. Alternatively the probe can be scanned over a defined area and electrons of a specified energy can be collected and displayed on a CRT to show the distribution of the corresponding element over the scanned area – precisely as X-ray maps are generated in SEM and STEM. The analyser consists either of a cylindrical mirror analyser or a hemi-cylindrical analyser. With an axial cylindrical analyser the collection efficiency of Auger electrons can be as good as 10%. To obtain information across a selected energy range, the potential across the analyser is ramped so that the energy of the particular Auger electrons which pass through the aperture onto the electron detector (a scintillator–photomultiplier or electron multiplier) varies with time.

The aperture situated in front of the detector can be varied in size and thus controls the energy resolution of the spectrometer, which can be varied between about 1 and 40 eV.

The range over which the scan control is ramped defines the energy range of the Auger electrons which are detected. Clearly with an unknown sample the maximum energy range (typically 0–2 kV) is scanned initially to determine which elements are present. The scan time can be increased so that the signal-to-noise ratio is improved. In virtually all Auger work the signal is differentiated to improve the detectability of Auger peaks. At small signal levels, single electron counting is used and at higher count rates a voltage to frequency converter can be used. The combination of single electron detection and voltage to frequency conversion allows a very large dynamic range, i.e. very low and very high count rates covering a range between 1 and about 10^9 counts per second.

The lateral resolution of an Auger signal is determined by the spreading of the electron beam as it passes into the sample since backscattered electrons can generate Auger electrons as they leave the sample. Improved lateral resolution can be obtained either by using as low an accelerating voltage as possible or by using very thin samples. The depth resolution is controlled by the energy of the Auger electrons; low energy electrons can come from only the top two or three atomic layers as Auger electrons produced at greater depths will be absorbed within the sample (see Chapter 6).

As shown in Fig. 3.6 an ion gun is an integral part of the equipment. This gun is used both to clean the samples after the spectrometer is pumped down to the high vacuum ($\sim 10^{-10}$ Torr) which is desirable for surface analysis work, and to remove the outermost layers of material so that the chemistry of the underlying material can be obtained by subsequent Auger analysis. In order to relate milling time to the amount of material removed it is necessary to make use of measured sputtering efficiencies [3]. These sputtering efficiencies are influenced by the chemical state of the atoms being sputtered and appropriate data are not always available. An alternative way to carry out depth profiling is to create a tapered section of the surface by producing a shallow recess in the specimen surface using a large diameter spherical-nosed drill. Auger spectra taken from the edge to the centre of the recess enable a depth profile to be obtained with a depth resolution defined both by the probe size and the radius of curvature of the spherical recess. Clearly care must be taken to minimize any changes introduced by the machining operation.

Many Auger spectrometers allow *in situ* fracture studies to be carried out so that the fracture surface is never exposed to air. This is clearly a very valuable technique for detecting the influence of segregation on fracture properties and a great deal of work has been done in this field [4]. Much of this work involves depth profiling but there are obvious difficulties associated with analysing and with profiling rough surfaces or surfaces which may be very inhomogeneous chemically. The details of the depth profiles should be treated with some caution.

Although early Auger spectrometers tended to be dedicated instruments there has been a tendency to develop combined Auger facilities which can operate as a high resolution SEM equipped with EDX together with ESCA and SIMS. As pointed out earlier in this chapter ESCA and SIMS are outside the scope of this book and the reader is referred to specialist texts covering these techniques [5].

An essential part of a modern Auger spectrometer is the ability to transfer specimens rapidly to the high vacuum chamber; a multisample carousel together with a rapid entry transfer system is essential. As with all the equipment discussed in this book the use of dedicated microprocessors and computers is central to the successful application to materials problems.

3.5 ELECTRON MICROPROBE ANALYSIS

Electron microprobe analysis (EMPA) is an established technique [6] used for rapid analysis of samples at a spatial resolution of about 2–3 μm. The layout of a modern microprobe is essentially similar to that of a SEM and indeed some instruments are hybrid microprobes/SEMs. A microprobe consists of an electron gun, a probe-forming system, an imaging system (which may be simply an optical microscope) and an X-ray detecting system.

The electron gun operates typically at voltages up to 30 kV. The condenser system is used to produce a small probe between 20 nm and 1 μm in size on the sample with currents of about 1 and 100 nA, respectively. If larger currents are required LaB_6 filaments (or even field emission guns) may be used but there is little advantage if the microprobe is to be used solely as a microprobe on bulk samples. Thus, as already discussed, the spatial resolution of X-ray analysis from bulk samples is limited to about 2 μm by spreading of the electron beam within the sample. The sample can be imaged with the emitted secondary electrons and the field of view to be analysed can therefore be selected precisely as in SEM or STEM.

Microprobes virtually always use wavelength dispersive X-ray (WDX) spectrometers (although the tendency now is to incorporate energy dispersive (EDX) as well) because of their superior resolution and because of the ability to detect X-rays from elements down to beryllium. The fact that WDX spectrometers can accept only a single wavelength X-ray at any one time leads naturally to the reqirement of two, or even three WDX spectrometers on a microprobe so that simultaneous analyses can be carried out. These spectrometers are mounted so as to maintain the correct geometry for X-ray collection so that fairly complex mechanisms are required (see Section 3.6).

Computer programs are available to obtain compositions from the relative intensities of X-ray lines but in view of the uncertainties in many of the parameters involved in the calculations it is commonplace to use standards in conjunction with the program if quantitative data are required. Thus the specimen stage in a microprobe is designed to hold several specimens and the appropriate standards. The intensities obtained from sample and standard can compared directly after correction for any drift in specimen current.

For quantitative analysis it is important that the sample be smooth so that the assumptions built into the computer program correcting for absorption, etc. are met with (see Chapter 6). With non-conducting specimens a conducting path must be provided and if polished samples are mounted in bakelite it is necessary to coat the mount (and non-conducting sample) with carbon in a carbon evaporator, but some care is necessary to ensure that a constant covering of carbon is used; if too much is used significant energy loss of electrons and absorption of low energy X-rays will take place.

Precisely as in SEM or STEM an X-ray map or X-ray line trace can be produced by using the output from an X-ray detector to modulate the image on the scanned image display. Both because very thin windows (and windlowless detectors) have been developed for the X-ray detector used in conjunction with WDX spectrometers, and because the sensitivity of WDX spectrometers is superior to that of EDX, it is possible to detect elements down to beryllium in the periodic table and thus produce distribution maps of most important elements in materials. The use of dedicated computers has resulted in a large increase in the efficiency of use of microprobes. For example the microprobe can be instructed to take successive analyses at a specific number of fixed points and to process the data using the appropriate software.

3.6 X-RAY SPECTROMETERS

Two types of X-ray spectrometer are used in electron beam equipment, wavelength dispersive (WDX) and energy dispersive (EDX). The aim of both types, of spectrometer is to obtain, in a useful form, the integrated intensities of the X-rays emitted from the sample. The separation of the various X-rays is achieved in a WDX system by using a crystal spectrometer and in an EDX system by using a silicon–lithium detector. These two systems will be dealt with in turn.

3.6.1 WDX spectrometer

For a crystal of known lattice spacing d, X-rays of a specific wavelength λ will be diffracted at an angle θ given by the well-known Bragg equation,

$$n\lambda = 2d \sin \theta \qquad (3.1)$$

Table 3.1 d spacings of diffracting planes and corresponding range of wavelength for some commonly used crystals in crystal spectrometers.

Material	d spacing of diffracting planes (Å)	Wavelength range (Å)
Lithium fluoride, LiF	2.013	1 – 3.5
Quartz, SiO_2	3.343	1.8– 5.5
Pentaerythirol, $C_5H_{12}O_4$	4.371	2.2– 7.0
Ammonium dihydrogen phosphate, $NH_4H_2PO_4$	5.32	2.8– 9.5
Mica, $KAl_3(SiO_4)_3$	9.92	5.0–17.5
Rubidium hydrogen phthalate, $C_8H_5ORb_6$	13.06	5.8–25

Different wavelengths are selected by changing θ and in order to cover the necessary range of wavelengths several crystals of different d spacings can be used successively in a spectrometer. Wavelengths required range from 1 Å to 25 Å; corresponding d spacings for practicable values of θ, which lie between about 15° and 65°, are shown in Table 3.1 for some commonly used crystals.

Crystals used in WDX spectrometers generally have curved rather than flat surfaces, in order to increase the efficiency of focussing the Bragg diffracted X-rays into the X-ray detector. Fig. 3.7 illustrates a typical geometry for a WDX spectrometer where the specimen (which is the X-ray source), the crystal and the detector all lie on a circle, the Rowland circle. If the radius of the Rowland circle is r then the crystal is bent to a curvature of $2r$. Different wavelength X-rays are collected by the detector by setting the crystal at different angles, θ. A 2:1 gearing is needed so that the detector moves through twice the angle that the crystal moves through in order to maintain the

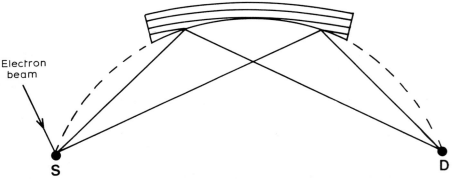

Fig. 3.7 Schematic ray diagram for a wavelength dispersive X-ray spectrometer showing the geometry for a focusing spectrometer. The specimen S generates X-ray signals which will be detected at D after being Bragg diffracted.

correct geometry. A fully focussing spectrometer requires that the specimen to crystal distance varies as $\sin \theta$ and the complicated mechanism which is required to maintain this geometry is simplified, with only a small loss in focussing, in several spectrometers [6]. The important property of a WDX spectrometer is that it acts as a monochromator, focussing the individual wavelength X-rays onto the detector. Clearly only one particular wavelength X-ray can be focussed onto the detector at any one time and this makes the operation of a WDX spectrometer time-consuming, especially if the elements present in the sample are unknown.

The resolution of WDX spectrometers is controlled in the main by the perfection of the crystal, which influences the range of wavelengths over which the Bragg condition is satisfied, and by the size of the entrance slit to the X-ray detector. If the perfection of the crystal is improved and/or the slit made narrower resolution can be improved at the expense of intensity and it is not meaningful to define the resolution without defining the other parameters. Nevertheless, for fairly standard conditions the resolution ($\Delta \lambda$) is about 0.01 Å which gives a value for $\lambda/\Delta\lambda$ of about 300. This leads to a peak-height: background ratio of about 250 for a medium atomic weight sample.

The X-ray detector used in conjunction with a crystal spectrometer is usually a proportional counter, which produces an electrical signal because the incident X-rays cause ionization of the gas in the counter. The electrons generated by the ionization are attracted to an anode wire in the counter and the signal generated by these electrons is amplified electronically. The resultant electrical signal is proportional to the X-ray energy (i.e. inversely proportional to the wavelength). The window of the counter must be thin and of low atomic number in order to reduce the amount of X-ray absorption. The output pulse from the counter is amplified and differentiated to produce a short pulse. The time constant of the electrical circuit is of the order of 1 μs which leads to possible count rates of at least 10^5 s^{-1}.

3.6.2 EDX spectrometers

EDX detectors have developed rapidly in the last ten years and have replaced WDX detectors on transmission microscopes and are used together with WDX detectors on microprobes and on SEMs. A schematic diagram of a SiLi detector is shown in Fig. 3.8 (a). X-rays enter through the thin beryllium window and produce electron–hole pairs in the SiLi. Each electron–hole pair requires 3.8 eV, at the operating temperature of the detector, and the number of pairs produced by a photon of energy E_p is thus $E_p/3.8$. The charge produced by a typical X-ray photon is about 10^{-16} C and this is amplified by the field effect transistor (FET). Further amplification is provided by conventional transistor circuits. The height of an amplified, shaped pulse is then a measure of the energy of the incident X-ray photon.

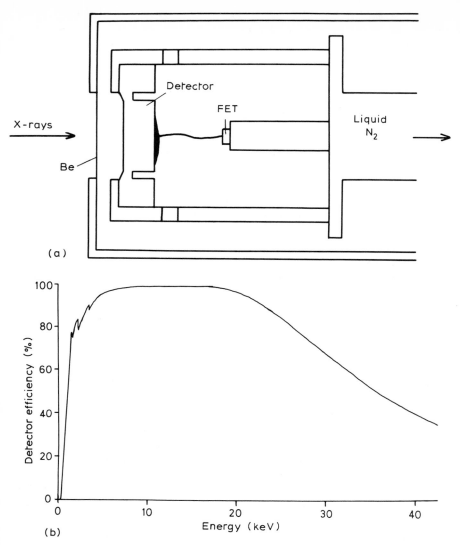

Fig. 3.8(a) Schematic diagram showing the geometry of a SiLi energy dispersive X-ray detector. (b) Diagram showing the efficiency of a SiLi detector as a function of X-ray energy.

The data are stored in a multichannel analyser. Given that the X-ray photons arrive with a sufficient time interval between them the energy of each incident photon can be measured and the output presented as an intensity versus energy display. The amplification and pulse shaping takes about 50 μs and if a second pulse arrives before the preceding pulse is processed, both pulses are rejected. This results in significant dead time for count rates greater than about 4000 s^{-1}. Since the count rate refers to the count across the whole energy range, typically 0–40 kV, the time for collecting statistically significant data from minor components can be prohibitively long.

The number of electron–hole pairs generated by an X-ray of a given energy is subject to normal statistical fluctuations and this taken together with electronic noise limits the energy resolution of a SiLi detector to about 160 eV at Mn_{K_α}.

The main advantage of EDX detectors is that simultaneous collection of the whole range of X-rays is possible and an indication (fingerprint) of all the elements (above sodium in the periodic table) can be obtained in a matter of seconds. The main disadvantages are: (1) the relatively poor resolution (of ~ 160 eV at Mn_{K_α}) which leads typically to a peak-height: background ratio of about 50; and (2) the limited total count rate.

3.7 ELECTRON SPECTROMETERS

As discussed in Chapter I electron spectrometers are an integral part of analytical transmission electron microscopes and of Auger facilities. The spectrometers which are commonly used for the two techniques are magnetic and electrostatic respectively and they will be discussed in turn.

3.7.1 Magnetic electron spectrometers

As mentioned in Chapter 1 magnetic prism electron spectrometers can be interfaced to both TEMs and to STEMs. The spectrometers which are interfaced to TEMs collect all the transmitted electrons lying within a cone of width α as shown in Fig. 1.6 and the intensity of the various electrons is obtained after dispersing the electrons with a magnetic prism, as discussed in Chapter 1. Because of the limited dispersive power and the aberrations of the magnetic spectrometer, the resolution depends on the object size δ at the spectrometer object plane and on the semidivergence angle β at the spectrometer object plane. As pointed out in Chapter 1 post-specimen lenses can be used as coupling lenses so that δ and β are respectively, but not identically, related to d the diameter of the specimen being analysed and α, the semiangle of the scattered electrons.

The coupling between the microscope and the electron spectrometer is best considered with reference to a particular design and the design chosen for illustration is the EM400/Gatan spectrometer. Two methods of coupling are used: image coupling (Fig. 3.9(a)) and diffraction coupling (Fig. 3.9(b)).

In diffraction coupling the microscope is operated in the normal TEM image mode with the region of interest centred on the viewing screen. This mode is termed diffraction coupling because, at the spectrometer object plane

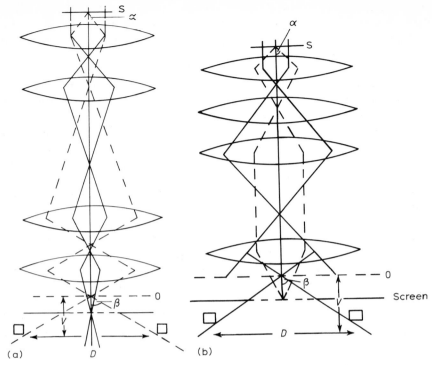

Fig. 3.9 Schematic ray diagram showing two methods of coupling an electron energy loss spectrometer to an electron microscope. (a) Image coupling: the spectrometer object plane is at a distance V from the entrance aperture D, to the spectrometer, which is below the TEM viewing screen. The collection semiangle of the spectrometer at the specimen is α. (b) Diffraction coupling: a diffraction pattern is formed at the object plane and an image of the selected area is visible on the final viewing screen. The area of the specimen sampled is defined by D.

(see Fig. 3.9), a diffraction pattern is formed by the final projector lens. The effective camera length for this diffraction pattern is given by Y/M where Y is the distance between the viewing screen and the spectrometer object plane and M is the magnification at the viewing screen. The acceptance angle of the spectrometer at the object plane is defined by D, the size of the entrance aperture to the spectrometer (which is situated about 23 cm below the viewing screen), and by the distance V. Thus $\beta = D/2V$ and for $V \sim 50$ cm (as on the EM400) and D between 0.1 and 0.5 cm, β lies in the range 1 to 5 mrad. Because the effective camera length at the object plane is so small, typically between 1 mm and 10 μm, the collection angle β is much smaller than α and it is thus possible to collect electrons scattered through relatively large angles* (up to 50 mrad) when β is only 2 mrad so that the resolution of the spectrometer is not degraded by aberration. This is a particularly simple EELS technique to use since the magnification on the viewing screen can be varied to define the area analysed, the size of which is given by D/M. At very high magnifications, corresponding to very small selected areas, the number of loss electrons entering the spectrometer will be very small and there will be signal-to-noise problems. In cases where very small regions are to be analysed it is preferable to use the STEM mode.

In the STEM mode and in the SAD mode a low magnification image is formed at the spectrometer object plane (hence image coupling) and a diffraction pattern is formed on the viewing screen (Fig. 3.9(a)). The size of the selected area is defined either by the probe size or by the STEM scan size. The collection angle at the spectrometer object plane is $D/2V$ and this is equivalent to a collection angle α at the specimen of $M_o D/2V$ where M_o is the magnification of the low magnification image at the spectrometer object plane. For $D = 0.1$ cm and $V = 50$ cm, $\beta = 1$ mrad and if $M_o = 10$ (a reasonable value) then $\alpha = 10$ mrad.

By the appropriate choice of D (typically ~ 2 mm) optimum coupling is obtainable for both of these modes of operation at a resolution of the order of 1 eV.

As mentioned in Chapter 1 the EELS can be used to produce an energy filtered diffraction pattern or STEM image. Energy filtered TEM images can be produced either by using an energy filtering microscope [7] or equivalently using a conventional TEM and post-spectrometer lenses.

3.7.2 Auger electron spectrometers

Although detailed layouts of the many commercial Auger spectrometers differ there are essentially only two different arrangements for collecting Auger electrons. One of these is based on a hemispherical analyser (HSA) and the other on a cylindrical mirror analyser (CMA) (cf. Fig. 1.7).

In spectrometers using HSA a wide angle lens is used to collect and focus the Auger electrons onto the entrance slit of the analyser. This lens is typically 5–10 cm from the specimen, allowing excellent access to the specimen. As discussed in Chapter 1 this lens is commonly used as a retarding lens and there are two different modes of operation of this retarding lens. In the first mode the energy of all the collected electrons is reduced by a factor, say twenty times, by operating the lens as a two electrode lens; the lens system and the analyser potential are then ramped so that the different incident electrons are collected with the increased resolution expected

---

* If the objective aperture is removed.

for a retarded lens system (see equation (1.14)). Alternatively the pass energy of the analyser is set at some preselected potential and the electrons collected by the lens are retarded to this preselected energy. This second method (fixed operating voltage for the analyser) requires that the potential of the middle electrode of the three electrode lens is scanned in a ramp, which is a non-linear function of the electron energy, in order to focus the electrons from the specimen onto the entrance slit to the analyser. This provides better sensitivity at low energies (because it avoids the problem of dealing with very low energies) but the first method is more useful for quantitative analysis since the only action of the lens is to retard the electrons (and hence improve the resolution) and the relative intensities of the electrons which enter the analyser accurately reflect the relative intensities of the Auger electrons from the same fixed area of the specimen.

In the CMA the analyser subtends a large angle at the specimen, because it is close. There are however two main disadvantages to the CMA concept: (1) the restricted working space associated with working distances of about 1 cm; (2) the optimum resolution and accurate values of energies are obtained only for samples positioned within a fraction of a millimetre of the focal point of the CMA. The first limitation can be overcome, with a corresponding loss of collection angle, by using a hemicylinder, rather than a cylinder. The analyser in a CMA consists of two concentric cylinders with entrance and exit apertures (the size of these apertures controlling the resolution which is typically 0.3% at 2.5 kV) and the potential between these cylinders is ramped in order to sweep the electrons over the exit aperture and hence onto the photomultiplier.

Generally the output is electronically differentiated to improve the visibility of small peaks and to remove the continuously sloping background (cf. Chapter 6).

REFERENCES

1. Hirsch, P.B., Howie, A., Nicholson, R.B., Pashley, D.W. and Whelan, M.J. (1965) *Electron Microscopy of Thin Crystals*, Butterworth, Sevenoaks.
2. Fraser, H.L., Loretto, M.H. and Jones, I.P. (1977) *Phil. Mag.*, **35**, 154.
3. Wehner, G.K. (1972) in *Methods in Surface Analysis* (ed. A.W. Czanderna), Elsevier, New York, p. 5.
4. Lea, C., Seah, M.P. and Hondros, E.D. (1980) *Mater. Sci. Engng*, **42**, 233.
5. Park, R.L., Houston, J.E. and Schreiner, D.G. (1970) *Rev. Sci. Instrum.*, **41**, 1810.
6. Reed, S.J.B. (1975) *Electron Probe Microanalysis*, Cambridge University Press, Cambridge.
7. Castaing, R. and Henry, L. (1962) *C.R. Acad. Sci. Paris*, **255**, 76.

4

INTERPRETATION OF DIFFRACTION INFORMATION

4.1 INTRODUCTION

The interpretation of diffraction information generated by electron–specimen interaction is discussed in this chapter and in Appendix A. The generation and collection of diffraction data have been discussed in earlier chapters, but the present interest lies in the information which can be obtained concerning the specimen; the aim of electron beam techniques being to obtain detailed structural and chemical information from defined regions of the specimen. It should be clear that many of the signals discussed in Chapters 4–6 can be obtained from the same defined region of the specimen and this is one of the most important factors underlying the success of electron beam instruments.

4.2 ANALYSIS OF ELECTRON DIFFRACTION PATTERNS

An electron diffraction pattern is, by definition, the angular variation of elastically and inelastically scattered electrons at an infinite distance from the crystal, i.e. electrons scattered in a specific direction may originate from anywhere in the specimen defined either by the selected area aperture* or by the probe in TEM or by the small probe in STEM. In a conventional TEM the objective lens focusses the scattered electrons in the back focal plane and a diffraction pattern is visible if this plane is imaged.

Many factors influence the form of a diffraction pattern. Thus, perfect crystals give rise to strongly diffracted beams in certain well defined and predictable directions and hence give rise to sharp diffraction maxima if a parallel beam of electrons is used as an incident beam; increasing the incident beam convergence angle results in discs of intensity in the back focal plane. The diameter of the discs defines the incident beam convergence as discussed later (see Section 4.2.2(c)). Under the appropriate conditions these discs contain detailed information from which point groups and space groups may be deduced.

In the following sections the interpretation of conventional diffraction

---

* Note the errors in area selection using SAD in Chapter 3.

patterns (i.e. those with sharp diffraction maxima) will be dealt with first, before the interpretation of convergent beam diffraction patterns, the interpretation of diffuse scattering, and of channelling patterns and electron backscattered patterns are discussed.

4.2.1 Conventional transmission electron diffraction patterns

The information available in diffraction patterns can be used to determine the crystal system and lattice type (although this can be difficult as will become apparent in this chapter) to which the specimen belongs; it can be used to provide the crystallographic information which is essential to the interpretation of defect images, i.e. to determine the electron beam direction, the indices of any operating reflection and the deviation from the exact diffracting condition for any reflection. It can be used to determine misorientations between crystals or orientation relationships between phases and it can be used to define the degree of order in an alloy.

The determination of the degree or order is not straightforward because, as pointed out in Chapter 2, the relative intensities of diffraction maxima will not be accurately reflected by the relative intensities derived from structure factor calculations (equation (2.4)). Thus, in contrast to X-ray diffractometry where the relative intensity of a superlattice and a fundamental reflection can be used immediately to determine the degree of order, the relative intensities of diffracted beams in electron diffraction patterns are strongly influenced by rediffraction. The amount of rediffraction is influenced by the precise electron beam direction the thickness of the sample, and the extinction distance of the excited reflections.

It is possible, however, to calibrate electron patterns so that the degree of order can be obtained by measuring the intensities of superlattice and fundamental reflections if the electron beam direction and specimen thickness are accurately known. Thus, if samples of identical thickness, for which the degree of order is known from X-ray work, are examined in identical beam directions then the ratio of intensities of fundamental and superlattice reflections can be used to determine the degree of order in a sample with an unknown degree of order if the thickness and beam direction are kept constant. As discussed in this chapter it is possible to set the beam direction very reproducibly using convergent beam patterns or Kikuchi lines and thickness can be measured within about 5%. Further discussion of the assessment of order using electron diffraction is postponed until Section 4.3.

(a) Solution of conventional diffraction patterns from crystals of known structure

A typical diffraction pattern obtained using the selected area technique (see Chapter 3) from a very thin crystal of molybdenum is shown in Fig. 4.1.

INTERPRETATION OF DIFFRACTION INFORMATION 67

Fig. 4.1 Selected area transmission electron diffraction pattern obtained at 100 kV from a very thin specimen of molybdenum. Directly transmitted beam marked O, and two diffracted beams X and Y.

The diffraction pattern consists of a series of well defined maxima; the most intense beam is the directly transmitted beam, i.e. a beam travelling in the same direction as the incident electron beam. The other maxima define the directions in which strong diffracted rays left the crystal.

The conditions which give rise to strong diffraction are defined by the Bragg equation

$$\lambda = 2d \sin\theta \qquad (4.1)$$

where λ is the wavelength of the electrons, d the interplanar spacing of the diffracting planes, and θ is the angle of incidence of the electrons on those planes. Since λ is small (0.037 Å for 100 kV electrons) the angle θ is about $1/2°$ for low-order planes. Strong reflections will be excited even when the incident beam does not satisfy equation (4.1) exactly, both because the specimens from which the patterns are obtained are thin (see [1], [2] and Appendix A) and less importantly, because the incident beam has a significant divergence. Planes giving rise to strong maxima are approximately parallel to the electron beam direction and when the electron beam is parallel to an important zone axis there will be a large number of planes sufficiently close to their Bragg angles to give rise to a large number of diffracted beams. The zone axis, i.e. the electron beam direction **B** referred to the crystal axes, can thus be obtained simply by determining the angle through which the electrons have been scattered for any two of the diffraction maxima. The angles through which the electrons have been scattered define d_{hkl} (from equation (4.1)) and hence the indices of the planes, since the interplanar spacings can be calculated using the appropriate formula (see Appendix B). The electron beam direction can be obtained from the indices of the planes by taking the cross product of their indices – as shown below. Before solving examples of diffraction patterns it is useful to discuss the Ewald sphere construction since this provides a simple way of considering electron diffraction patterns. This construction, which is formally equivalent to Bragg's law, is carried out as follows.

A line OX, of length $1/\lambda$, is drawn through the origin O of the reciprocal lattice* in the direction of the incident beam. A sphere of radius $1/\lambda$ is described about X and if any reciprocal lattice point P, other than the origin

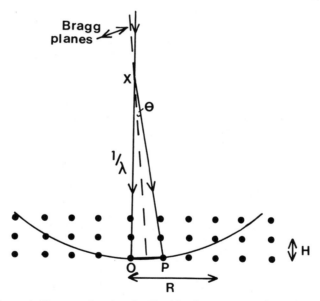

Fig. 4.2 Schematic diagram showing the Ewald sphere construction. The planes which are at the Bragg angle θ are marked. O is the origin of the reciprocal lattice, P a point on the Ewald sphere which is centred at X. Three layers of reciprocal lattice points are shown and the intersection of the sphere with points in these zones can be seen. These high order Laue zones are discussed in Section 4.2.2(b). The radius of the first zone is R and the spacing along OX of these layer planes is H.

* See Appendix A for the definition, some properties, and derivation of a reciprocal lattice.

O, is intersected by the sphere then a diffracted ray is generated in the direction XP. It is easy to show from Fig. 4.2 that the construction is equivalent to Bragg's law.

Since $1/\lambda \gg 1/d$ it is clear that a diffraction pattern can be viewed as a planar section through the origin of the reciprocal lattice*. **B** can therefore be determined if we can index any two reciprocal lattice vectors such as OY and OX as $[h_1^* k_1^* l_1^*]$ and $[h_2^* k_2^* l_2^*]$ so that $\mathbf{B} = [h_1^* k_1^* l_1^*] \wedge [h_2^* k_2^* l_2^*]$. In this book the cross product is taken in the sense to obtain **B** upwards from the photographic plate, which should be printed emulsion side up.

The indexing of diffraction maxima is generally carried out in one of two ways. In the first method the effective camera length L is either known or can be calibrated so that the measured distances OX, OY on a plate can be used to obtain the spacing of the planes which are giving rise to the maxima. It can be seen from simple geometry (for small angle) that $d_1 = \lambda L/|OX|$ and $d_2 = \lambda L/|OY|$. These values of interplanar spacings d_1 and d_2 enable the indices of the planes to be defined. The specific indices for these planes can be obtained from the angles between OX and OY as discussed below.

The second method uses the ratios of the distances OX and OY, rather than the absolute distances. Distances of diffraction maxima from the direct beam are proportional to the reciprocals of the spacings of the corresponding $\{hkl\}$ planes in real space (see Appendix B) and hence the ratio

Fig. 4.3 Selected area transmission electron diffraction pattern obtained from a very thin specimen of titanium. Directly transmitted beam marked O and two diffracted beams marked X and Y.

* If high-angle diffraction information is collected it is obvious that the section is not planar. This aspect is dealt with in Section 4.2.2.

Table 4.1 Measurements from the diffraction pattern from molybdenum in Fig. 4.1 and the derivation of **B**, the electron beam direction, from the indices of the two reflections X and Y.

Measured distances	Squares of OX and OY	Ratios of squares	Possible indices	Specific indices	Beam direction **B**
OX = 1.7	2.89	2	200 $(h^2 + k^2 + l^2) = 4$	200	[011]
OY = 1.2	1.44	1	110 $(h^2 + k^2 + l^2) = 2$	0$\bar{1}$1	

of the squares of the magnitudes OY and OX gives the ratios of squares of the interplanar spacings. This ratio together with the measured angle between the maxima (the angles between the planes $(h_1 k_1 l_1)$ and $(h_2 k_2 l_2)$ in the real crystal) enables **B** to be obtained.

Two diffraction patterns will be solved to illustrate the methods outlined above. The first, shown in Fig. 4.1 is taken from a b.c.c. sample, molybdenum. The second, Fig. 4.3 is taken from an hexagonal close packed (h.c.p.) sample, titanium.

In Fig. 4.1 the measured distances OX and OY are 1.7 and 1.2 cm respectively. As shown in Table 4.1, the ratio of the squares of these distances is as 2 is to 1 so that using the ratio method, possible indices are 200 and 110 for the reflections X and Y. Similarly using the known camera length for this pattern the d spacings were found to be 1.6 and 2.2 Å respectively, which correspond to the spacing of {200} planes and {110} planes, using the relationship*

$$d_{hkl} = \frac{a}{(h^2 + k^2 + l^2)^{1/2}} \quad (4.2)$$

* It should be noted that some confusion exists in the literature concerning this and the other equations for interplanar spacing given in Appendix B. In most textbooks dealing with crystallography, Miller indices are reduced to the smallest whole numbers so that, for example (110) refers to *all* planes parallel with the plane intersecting the x and y axes at one unit distance from origin. If this convention is used then it is essential to note that the equations for interplanar spacing refer only to the simple cells. For non-primitive crystals (e.g. b.c.c., f.c.c., h.c.p.) the atoms which are *not* at the corners of the unit cells give rise to planes which are parallel with, and at a smaller spacing than, the atomic planes which are present in the primitive cells. If this method of indexing planes is used these new interleaving planes have identical indices to those parallel planes in the primitive cell and the equations in Appendix B will not give the correct spacing for these planes. For example in an f.c.c. crystal all planes for which h, k and l are not all even or odd will have spacings one half of that given by these equations.

A preferable approach, implicit in work involving the reciprocal lattice, is to allow indices of reflections to have common factors and to allow the corresponding planes – which need *not* pass though atoms – to have indices with common factors. For example, 200 reflections, in molybdenum originate from (200) planes with half the spacing of the (100) planes and the spacing of the (200) planes is given correctly by equation (4.2).

INTERPRETATION OF DIFFRACTION INFORMATION

where hkl are the Miller indices of the plane and a the lattice parameter.

In order to assign specific indices to these two reflections it is necessary to ensure that the indices are such that the measured angle $\widehat{XOY}$ corresponds to that between planes of the assigned Miller indices. In the present case $\widehat{XOY}$ is 90° and thus 200 and $0\bar{1}1$ are possible indices for the two reflection X and Y. The cross product of these gives [011] as **B**. The solution of [011] obtained for **B** for the pattern shown in Fig. 4.1 is arbitrary in the sense that a different, but equivalent choice of indices for OX and OY would yield [101], [110], ... Specific indexing of the first reflection is totally arbitrary, indexing of the second is then constrained somewhat because the choice must be consistent with the measured angle between OX and OY. The indexing of **B** is then fixed and as will be discussed later the indexing of any subsequent diffraction patterns obtained after tilting the specimen is totally fixed by the indexing of the first pattern.

Solutions of electron diffraction patterns are particularly straightforward for cubic materials since the cubic symmetry is retained in the reciprocal lattice (see Appendix A). The reciprocal and real axes are collinear and therefore $[hkl]$ is parallel with $[h*k*l*]$ since by definition $[h*k*l*]$ is perpendicular to (hkl), and in a cubic system $[hkl]$ is perpendicular to (hkl). These simplifications arise in a cubic crystal because the unit translations along the three axes are equal and orthogonal. This is not true for other crystal systems and care must be taken not to carry over these simplifications into other crystal systems.

A similar procedure to that illustrated in Table 4.1 can nevertheless be carried out for crystals of lower symmetry and an example for an h.c.p. crystal is given in Fig. 4.3 from titanium and the corresponding calculations are shown in Table 4.2. Again by definition the magnitudes of the vectors OX and OY are inversely proportional to the corresponding interplanar spacing of the crystal, i.e.

$$|\mathbf{g}|^2 = \left(\frac{1}{d}\right)^2 = \frac{4}{3}\left(\frac{h^2 + hk + k^2}{a^2}\right) + \left(\frac{l}{c}\right)^2 \tag{4.3}$$

Table 4.2 Measurements from the diffraction pattern from titanium in Fig. 4.3 and the derivation of **B**, the electron beam direction, from the indices of the two reflections X and Y.

Measured distances	Ratios of distances	Possible indices	Specific indices	Beam direction **B**
OX = 1.6	1.0	$10\bar{1}0$	$10\bar{1}0$	$[1\bar{2}13]$
OY = 1.8	1.13	$\bar{1}101$	$01\bar{1}1$	

where the values of c and a appropriate to the specimen must be known (cf. Appendix B). Hence the measured distances on the plate are used to ascribe indices to the diffraction maxima. In the case of crystals of non-cubic symmetry it is generally worth calibrating distances on the plate so that they can be converted instantly into interplanar spacings and then using tables of interplanar spacings to identify the diffraction maxima. Again the measured angles on the diffraction pattern must be used to obtain internally consistent indexing.

If the magnitudes of the vectors in Fig. 4.3 are considered then possible indices for OX and OY are $\bar{1}101$ and $10\bar{1}0$ and the angle $\widehat{XOY}$ shows that the specific indices $01\bar{1}1$ and $10\bar{1}0$ would be appropriate. To take the cross product of these Miller Bravais indices and hence to obtain **B** in Miller Bravais indices the following procedure can be used. Because the indices of the reflections are the indices of planes in the real crystal the Miller indices can be obtained by omitting the third index so that $[01\bar{1}1]^*$ becomes $[011]^*$ and $[10\bar{1}0]^*$ becomes $[100]^*$. The cross product of $[100]^* \wedge [011]^*$ gives $[0\bar{1}1]$ as the Miller indices $[uvw]$ of **B**. In order to return to Miller Bravais indices, $[hkil]$, the following relationships which relate $[uvw]$ and $[hkil]$ are used: $h = 1/3[2u - v], k = 1/3[2v - u], i = -[h + k]$ and $l = w$. On this basis $[0\bar{1}1]$ is equivalent to $[1\bar{2}13]$.

Clearly non-cubic crystals are more difficult to deal with, but if tables of interplanar spacing, angles between planes, and angles between directions are available, solutions can usually be found fairly quickly.

The solution for **B**, obtained simply from the diffraction maxima, is inherently inaccurate because the diffraction maxima are not infinitely small points but are in fact rods along a direction parallel to the thin direction of the crystal (Appendix A). This extension is so large that diffraction maxima whose centres lie well away from the Ewald sphere can give rise to significant intensity if the rods are long enough to intersect the Ewald sphere. (For details see [1], [2] and Appendix A.) Errors in **B** of up to 15° can be caused and, in crystals which are thick enough for significant inelastic scattering to occur, use is made of the Kikuchi lines, which these electrons generate, in order to determine **B** accurately. The following section discusses this and other aspects of Kikuchi lines.

(b) Indexing and information available from Kikuchi lines

As can be seen clearly in the diffraction pattern shown in Fig. 4.4(a) there are sets of lines (called Kikuchi lines) which have the identical spacing to the distance of the spots from the direct beam.

The Kikuchi lines are formed by Bragg diffraction of electrons which have been previously inelastically scattered. These lines are an invaluable aid to the experimentalist, because not only do they allow an accurate determination of **B**, but they enable the exact deviation from the Bragg

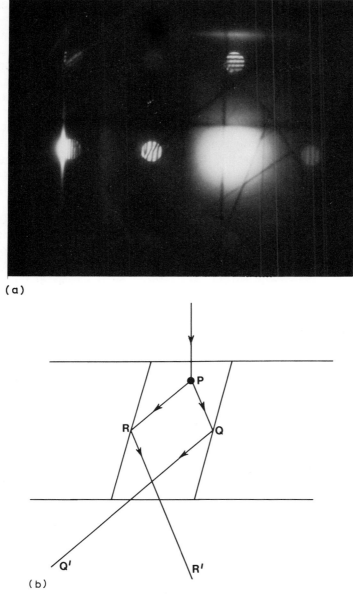

Fig. 4.4 (a) Selected area transmission diffraction pattern obtained from a sample of molybdenum thick enough to show Kikuchi lines. (A small probe was used to select the area. Hence the large diameter of the diffraction maxima—see Fig. 4.8 and text.) (b) Schematic diagram showing the origin of Kikuchi lines from inelastically scattered electrons, generated at P, which are Bragg diffracted by the planes indicated.

condition to be defined and enable controlled tilting to be carried out, about defined axes, to preselected electron beam directions.

A simplified explanation for the origin of Kikuchi lines is illustrated in Fig. 4.4(b). Electrons inelastically scattered at P are Bragg diffracted by the planes marked. Because P is a spherical source of electrons, with intensity peaked in the forward direction (see Chapter 2), the Bragg scattering of the inelastically scattered electrons results in the intensity in the direction QQ' being greater than that along RR'. Pairs of Kikuchi lines will be generated for all sets of planes, but will be visible (and intense) only for those planes which are sufficiently nearly vertical for the electrons to be imaged by subsequent lenses below the objective lens. If Fig. 4.4(b) is drawn so that the one set of planes marked are at the Bragg angle for the incident electrons the line QQ' will pass through the diffracted beam and RR' through the direct beam.

This simplified version generally defines the position of the Kikuchi lines very well but does not define either their intensities or their behaviour as a function of voltage. Many-beam calculations have been carried out [3] and the detailed behaviour modelled, but for our purposes the value of Kikuchi lines can be discussed satisfactorily on the basis of this simple model.

The positions of the Kikuchi lines on diffraction patterns are controlled only by the angle which the diffracting planes make with the incident beam. At the Bragg condition the lines must pass through the direct beam and through the appropriate diffraction maximum. As the crystal is tilted, so as to change the angle between the incident beam and these planes, the Kikuchi lines act as if the planes are mirrors, reflecting those inelastically scattered electrons which are now at the Bragg angle to these planes. Hence the Kikuchi lines move in a sense that is controlled by the sense of tilt of the crystal. Note that in contrast the spots may move, but only fractionally, before falling to zero intensity as the crystal is tilted so that the Ewald sphere no longer cuts the diffuse maxima around the reciprocal lattice points. Because all planes give rise to Kikuchi lines, the pattern of lines visible on the final screen changes as the specimen is tilted and the viewing screen is intersected by different sets of Kikuchi lines.

Fig. 4.5 shows two diffraction patterns, the second of which was taken after tilting the crystal through the [011] beam direction about the axis indicated. It is clear that a part of Kikuchi space is common and these diffraction patterns illustrate the changes which can be observed continuously during tilting. The manner in which the Kikuchi lines change during tilting defines the sense and enables the magnitude of the tilt to be observed, monitored and indexed. Schematic Kikuchi maps are illustrated in Appendix C. The maps are an invaluable aid in selecting tilt axes when specific **B**'s are required. Precise electron beam directions can be obtained

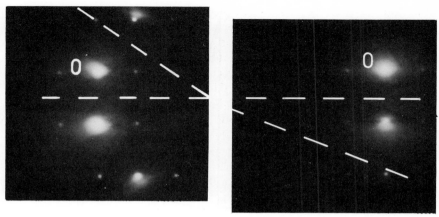

Fig. 4.5 Two transmission electron diffraction patterns taken either side of **B** = [011] demonstrating the change in the Kikuchi pattern on tilting through [011] about 200. The precise position of [011] is indicated by the intersection of the bisectors of pairs of Kikuchi lines.

simply by measuring the distance on the pattern (and hence the angle since the known Bragg angles for the reflections enable distances to be converted to angles) from a recognizable pole $[hkl]$ and the beam direction $[h_1 k_1 l_1]$ can then be specified as a known angle from $[hkl]$ about a defined tilt axis. Thus on Fig. 4.5(b) the electron beam direction is about 3° from [011] in a sense towards [012]. For the material scientist the observation of 20° of reciprocal space on a single pattern coupled with ±60° double tilting stages enables rapid identification of **B** and selection of virtually any desired beam direction and diffraction condition.

In addition to determining **B** and defining tilt axes, Kikuchi lines are used to determine the deviation parameter **s**. This parameter, which is a measure of the deviation from the Bragg condition of a reflection, is very important in influencing the contrast observed from crystal defects, as will be discussed in Chapter 5. The value of **s** is zero when the appropriate excess Kikuchi line passes through **g** and the corresponding deficit line through the direct beam and is taken arbitrarily to be negative when $\theta < \theta_B$. If we consider the situation at symmetry (see Fig. 4.6) we can see that $|s| = -g^2 \lambda/2$ and the value for any other position of the Kikuchi lines can be calculated by simple proportion. Because first order Kikuchi lines are commonly very diffuse it is better either to use the second order lines to define the position of the first order lines or to calculate s_g directly from them.

On the above basis it can be seen that it is straightforward to determine and select **B**, which is the projection direction in which the crystal is imaged in the corresponding micrograph; it is possible to select and define $\mathbf{g}_{hkl}$ the indices of any diffraction maximum, and as is evident from Fig. 4.5, it is

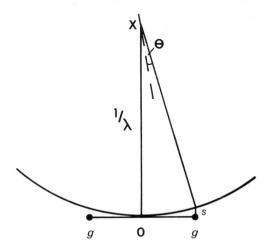

Fig. 4.6 Schematic diagram using the Ewald sphere construction illustrating the geometry at symmetry, from which the value of **s**, the deviation parameter, can be obtained for any orientation by simple proportion.

possible to tilt the sample so that only two strong beams appear on the diffraction pattern – the direct and the selected diffracted beams. It is also straightforward to determine s_g, the exact deviation from the Bragg condition. The three parameters **B**, **g** and s_g are required for interpretation of defect images, as will be discussed in Chapter 5.

(c) Determination of the crystal system from conventional electron diffraction patterns

The strong scattering of electrons by crystals leads, in all but the thinnest crystals, to rediffraction so that care is needed if the relative intensities of the diffraction maxima are to be used to obtain structural information. The ability to tilt in a controlled way about any axis does, however, enable the crystal system and unit cell dimensions to be determined from conventional electron diffraction patterns. The way this is done is best illustrated with an example.

The three schematic diffraction patterns shown in Fig. 4.7 were taken after tilting about the directions indicated by the Kikuchi lines in Fig. 4.7. The anlges of tilt between the three electron beam directions (**B**'s) are also marked in the figure; these angles were obtained in this case by two methods: firstly by noting the tilt reading at each of the three **B**'s and calculating the angle of tilt about the one axis to which the two tilts were equivalent; and secondly by taking diffraction patterns of the cubic matrix (in which this unkown phase was formed) at each of the three poles. Since the angles between the recognizable cubic **B**'s are known, the angles between the **B**'s in Fig. 4.7 are

INTERPRETATION OF DIFFRACTION INFORMATION 77

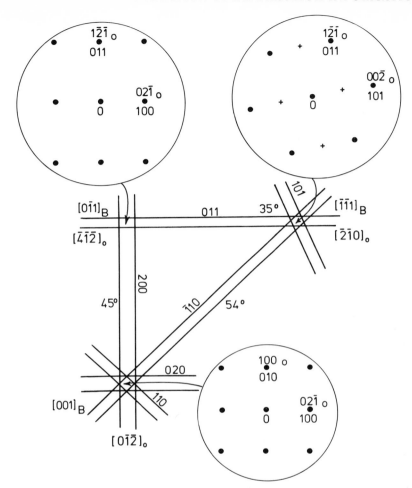

Fig. 4.7 Indexed schematic diffraction patterns taken from the same area of a second phase particle which was formed in a matrix which was cubic. The tilt axis and angles between the various patterns are indicated. (The subscripts o and B refer to orthorhombic and cubic phases respectively.) See text for discussion.

immediately obtained. The angles between **B**'s are angles between directions in the real crystal, the angles between diffraction maxima are angles between planes in the real crystal and the spacings of the diffraction maxima are inverse measures of the spacing of planes in the real crystal.

In order to determine the crystal system from these diffraction patterns some trial assumption must be made using information from just one of the patterns and then testing the deductions using the information from the other two patterns (note two patterns rather than three are required but the third pattern provides a double check on the conclusions). It is common sense to try to index the crystal as cubic, as a first assumption, and if that fails, to try successively more complex crystal systems since relatively tedious calculations are required to obtain interplanar angles and spacings. An examination of the symmetry of the patterns can save a large amount of calculation. For example, in the present case no poles were observed which showed either fourfold or threefold or sixfold symmetry. Hence the crystal is not cubic, tetragonal trigonal or hexagonal. Calculations based on the assumption that the phase was orthorhombic showed that the poles can be indexed as $[0\bar{1}2]$, $[\bar{4}\bar{1}2]$ and $[\bar{2}\bar{1}0]$ and that the material could be indexed as orthorhombic with $a = 3.252$ Å, $b = 9.734$ Å and $c = 4.155$ Å.

The present example can also be used to show how orientation relationships can be obtained from conventional diffraction patterns. Thus the $[0\bar{1}2]_o$ pattern was taken from the orthorhombic phase at the same tilt setting as those used for the $[001]$ diffraction pattern from the cubic matrix. The solution for **B** for these two patterns (using Kikuchi lines so that **B** is accurately defined) are precisely $[0\bar{1}2]_o$ and $[001]_{cubic}$ and these two directions are therefore parallel with each other in real space. The diffracting vectors $[100]^*_o$ and $[010]^*_{cubic}$ are also parallel in the two patterns so it follows that these two sets of planes are parallel. The apparent orientation relationship is

$$[0\bar{1}2]_o \| [001]_{cubic} \quad \text{and} \quad (100)_o \| (010)_{cubic}$$

It is useful to transfer the derived relationship to a stereogram in order to ensure that the simplest relationship is used to define the orientation relationship, and when this is done it is found that

$$(010)_{cubic} \| (100)_o \quad \text{and} \quad [101]_{cubic} \| [00\bar{1}]_o$$

One final practical point concerning tilting through large angles should be mentioned. This can be particularly difficult if the region of interest is very small because double-tilt stages are not doubly eucentric and the region of interest will move during tilting and its diffraction pattern disappear. This problem can be overcome to a significant extent, either by defocussing the diffraction pattern so that image information is visible within the direct and diffracted beams and any movement of the specimen can be compensated for with the specimen traverses, or by tilting in the image mode in dark field; the dark field being obtained by imaging with the **g** about which tilting is to be carried out. It is then straightforward to maintain the region of interest bright by tilting using both tilt controls as appropriate. The disadvantage of this latter method is that prominent poles can be missed and it is necessary to return to the diffraction mode periodically in order to check on this.

Clearly the above technique can be applied either with a selected area aperture to define the region of interest or with a fine probe. The use of a probe becomes essential if very small regions are to be characterzed so that a STEM pole piece is essential. If a small condenser aperture is used ($\sim 30\ \mu m$) then the beam convergence will be small (say 2×10^{-3} rad) and the diffraction pattern will be essentially identical to a selected area pattern. If, however, the convergence is made large by using a large ($\sim 150\ \mu m$) C2 aperture (to give a beam convergence $\sim 10^{-2}$ rad) then the diffraction pattern will consist of large discs which contain further information. These convergent beam diffraction patterns are discussed in detail in the following sections.

4.2.2 Convergent beam diffraction

A ray diagram illustrating the formation of a convergent beam diffraction pattern (CBDP) is shown in Fig. 4.8 and it is clear from this diagram that the

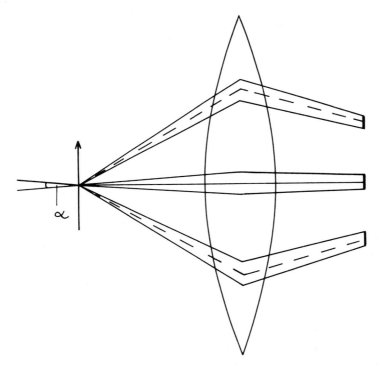

Fig. 4.8 Schematic ray diagram illustrating the formation of a convergent beam diffraction pattern in the back focal plane. The incident convergence angle defines the diameter of the convergent beam discs.

diameter of the diffraction maxima is defined by the beam convergence α. The discs of intensity which are formed in the back focal plane contain information concerning specimen thickness, orientation, point group and space group of the specimen. Much of this information is sensitive to small changes of specimen thickness and orientation, and in order to avoid averaging effects over a range of thickness or orientation it is necessary to use very small probes so that only a small volume of the sample will contribute to the diffraction pattern. Hence CBDPs are best obtained using microscopes which have STEM pole pieces.

The additional information available in CBDPs can be broadly divided into there types:

(1) Fringes within discs formed by strongly diffracted beams. If the crystal is tilted to two-beam conditions, these fringes can be used to determine thickness very accurately [4].
(2) High angle information in the form of fine lines which are visible in the direct beam and in the higher order Laue zones (HOLZ). These HOLZ are visible in a pattern covering a large enough angle in reciprocal space.
(3) Detailed structure, within the direct beam and within the diffracted beams, shows certain well defined symmetries when the diffraction pattern is taken precisely along an important zone axis [5].

The application of convergent beam diffraction to the determination of crystal thickness will be discussed in the first section below and the use of zone axis patterns in determining the symmetry elements of crystals will be discussed in the subsequent section.

(a) Determination of foil thickness using convergent beam diffraction

The calculated variation in intensity of a diffracted beam as a function of deviation from the Bragg condition for a crystal of thickness t is illustrated* in Fig. 5.2. Such rocking curves (cf. Fig. 4.9(b)) are precisely equivalent to the information contained in a disc of a CBDP obtained under two beam conditions as shown in Fig. 4.9(a).

As shown in Chapter 5, the intensity of the diffracted beam at the bottom of a foil is given by

$$I = \left(\frac{\pi}{\xi_g}\right)^2 \frac{\sin^2(\pi t s_{\text{eff}})}{(\pi s_{\text{eff}})^2} \tag{4.4}$$

where ξ_g is the extinction distance for the operative reflection and $s_{\text{eff}} = (s^2 + 1/\xi_g^2)^{1/2}$ and s is the deviation from the Bragg condition. Thus the positions of the minima are given by the condition $(t s_{\text{eff}})^2 =$ an integer, i.e.

---

* The calculation of this type of rocking curve is discussed in Section 5.3 for two beam diffraction conditions.

INTERPRETATION OF DIFFRACTION INFORMATION

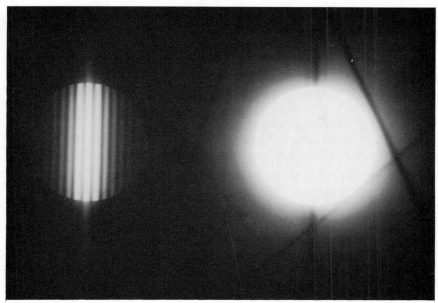

(a)

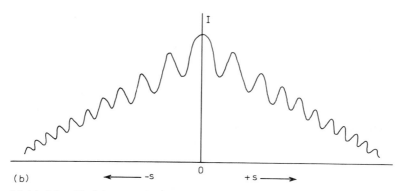

(b)

Fig. 4.9 (a) Magnified image of a convergent beam diffraction pattern taken from aluminium at 120 kV. The diffracted beam shows minima corresponding to those in (b). The precise positions of the minima depend on the foil thickness (cf. equation (4.5)).

by the condition

$$t^2 \left(s_i^2 + \frac{1}{\xi_g^2} \right) = n_i^2 \tag{4.5}$$

where s_i is the deviation of the ith minimum from the exact Bragg condition. Equation (4.5) can be rewritten [4] in the form

$$\left(\frac{s_i}{n_i}\right)^2 + \left(\frac{1}{n_i}\right)^2\left(\frac{1}{\xi_g}\right)^2 = \frac{1}{t^2} \qquad (4.6)$$

and plotting $(s_i/n_i)^2$ against $(1/n_i)^2$ gives the thickness (or more strictly $1/t^2$) as the intercept on the $(s_i/n_i)^2$ axis. The slope of the straight lines gives $(1/\xi_g)^2$ and hence ξ_g. In order to plot (s_i/n_i), values of s_i are needed and these are obtained by the technique discussed in Section 4.2.1(b). Thus if the distance $2\theta_B$ is measured on the diffraction pattern, the value of s_i, for the minima indicated on Fig. 4.9, is given by

$$s_i = \frac{g^2 \lambda}{2\theta_B}(\Delta x_i) \qquad (4.7)$$

where Δx_i is the distance of the ith minimum from the centre of the diffracted beam, i.e. from the condition $s = 0$.

Hence if **g** is identified and the values of λ and θ_B are known, then s_i for each minimum can be obtained from the appropriate Δx_i. The appropriate value of n_i is found by trying successively higher values of n_i until a good straight line is found which yields a reasonable* value of ξ_g. For a foil less than one extinction distance thick $n_i = 1$, for a foil between 1 and 2 extinction distances thick $n_i = 2$, etc.

Measurements taken from Fig. 4.9 yield a good straight line and give $t = 1330$ Å and $\xi_g = 1683$ Å (when $n = 1$) which is close to the calculated value of ξ_g for a 311 reflection in aluminium at 120 kV. Thicknesses with an accuracy of around 2% can be obtained in this way and this technique is very important in EDX work (see Chapter 6).

(b) Indexing of high order Laue zones

In previous sections of this book diffraction patterns have been interpreted as if they were planar sections through the origin of the reciprocal lattice. The radius of the Ewald sphere for 100 kV electrons is about 25 Å^{-1} and the spacing between reciprocal lattice planes about 1–2 Å^{-1}. The sphere therefore curves away significantly from the zero order Laue zone and, if the collection angle is large enough, the intersections of the sphere with the first and higher zones will be visible as circles of diffraction maxima centred on the direct beam when **B** is precisely parallel to $[uvw]$ (e.g. Fig. 4.10). These circular bands of maxima, known as high order Laue zones (HOLZ), are important because they provide three-dimensional information in a diffraction pattern.

---

* Low order reflections should be avoided since dynamical interactions along the systematic row not only give rise to significant changes in extinction distance, but also influence the position of the minima in the CBDP.

INTERPRETATION OF DIFFRACTION INFORMATION

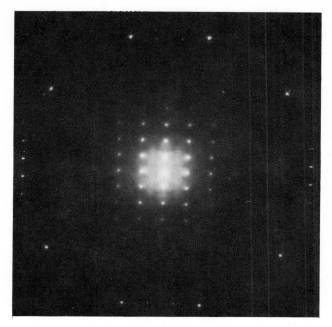

Fig. 4.10 Selected area transmission electron diffraction pattern taken from molybdenum zero and first order Laue zones.

The radius R_L of a first order Laue zone (FOLZ) is given by

$$R_L = \left(\frac{2H}{\lambda}\right)^{1/2} \quad (4.8)$$

where H is the spacing† of the $(uvw)^*$ planes along $[uvw]^*$, as is clear from the drawing of the Ewald sphere construction shown in Fig. 4.2. The fact that the interplanar spacing of $(uvw)^*$ along $[uvw]^*$ can be calculated from a diffraction pattern allows, for example, the presence of polytypes to be detected immediately and allows distinction between cubic and tetragonal patterns even when viewed along [001], so that in some cases the crystal system can be established from a single diffraction pattern.

The indexing of diffraction maxima in HOLZ can be done by first indexing the maxima in the zero order zone and then determining the magnitude of the reciprocal lattice vector $[uvw]^*$ separating $(uvw)^*$ planes along the beam direction $[uvw]$. The method is best understood by way of an example and

† As $[uvw]^*$ is written in a form without common factors (e.g. [110] rather than [220]) the spacing of $(uvw)^*$ will not always be given correctly by equations such as (4.2) (see footnote to (4.2)). Also, when considering non-primitive cells such as b.c.c. the size of the f.c.c. reciprocal lattice is $2/a$ and not $1/a$, where a is the edge length of the b.c.c. crystal (see Appendix A, p. 181). Both of these factors must be taken into account when calculating H.

(a)

(b)

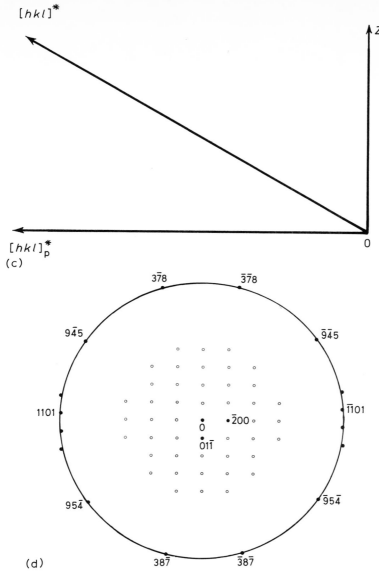

Fig. 4.11 (a) Indexed zero order layer corresponding to Fig. 4.10. The projected position of $[101]^*$ is marked as 101_p. (b) Indexed first order layer obtained by arbitrarily selecting $[101]^*$ as a vector which lies in the first zone (cf. equation (4.10)) and adding this vector to all those in (a). (c) Diagram illustrating the relation between $[hkl]^*$, a reciprocal lattice vector in the first order layer, and $[hkl]_p^*$, its projection onto the zero order layer and OZ the vector along $[uvw]$, the electron beam direction from O, the origin of the reciprocal lattice to Z where $[uvw]$ intersects the first order layer. (d) Zero and first order Laue zones oriented correctly by superimposing $[101]^*$ on its projected position (see (a)) in the zero layer. The drawn-in circle passes through the indexed first order maxima. This part of this figure should be compared with the first order zone in Fig. 4.10.

the diffraction pattern in Fig. 4.10 will be used. This pattern was taken from a molybdenum single crystal and examination of the zero order pattern shows that the electron beam direction is [011] (cf. Fig. 4.1) and that the diffraction maxima can be indexed as shown in Fig. 4.11(a). Effectively the electron beam direction $[uvw]$ is found from the zero order zone using the relationship

$$h^*u + k^*v + l^*w = 0 \qquad (4.9)$$

where $h^*k^*l^*$ is any reflection in the zero order pattern. Similarly for any higher order zone for any crystal system the reciprocal lattice plane $(uvw)^*$ is described by a relationship of the form

$$h^*u + k^*v + l^*w = N \qquad (4.10)$$

where N is an integer[†] and $h^*k^*l^*$ any reflection in the zone. In the example illustrated in Fig. 4.10 the value of $N = 1$ in equation (4.10) is satisfied, for example, for the 101 reflection and it is therefore possible to draw out the first order layer simply by adding 101 to each reciprocal lattice point in the zero layer as shown in Fig. 4.11(b). Note that some reflections may be absent in the zero layer or the first order layer and it is only the section of the reciprocal lattice which must be identical; the visible pattern of diffraction maxima may differ if structure factor considerations cause absences in one of the zones. The next step in indexing the maxima in the FOLZ is to position the first order zone with respect to the zero order zone since the maximum labelled 101 in Fig. 4.11(b) has been selected arbitrarily as one reflection satisfying equation (4.10). The actual direction of $[101]^*$ is of course already defined by the indexing of both the maxima in the zero order layer and **B**, and it is the direction of $[101]^*$ we need to know.

Thus if we take any point $[hkl]^*$ in the first layer and project this along the beam direction $[uvw]^*$ onto the zero layer (see Fig. 4.11(c)) to give the point $[hkl]^*_p$ then the vectors $[hkl]^*$ and $[hkl]^*_p$ differ by the separation of the (uvw) planes along $[uvw]^*$. This vector, OZ on Fig. 4.11(c), is given by $N[uvw]^*/(u^2 + v^2 + w^2)$. In the present example OZ is thus $\frac{1}{2}[011]^*$ and we can therefore write

$$[101]^*_p = [101]^* - \tfrac{1}{2}[011]^* = \tfrac{1}{2}[2\bar{1}1]^*. \qquad (4.11)$$

This point is indicated on Fig. 4.11(a). Using the fact that $[101]^*$ must lie vertically above $\frac{1}{2}[2\bar{1}1]$ in the zero layer the two layers have been superimposed in Fig. 4.11(d), and the indices of those lying on the circle in Fig.

[†] For different crystal structures the possible values of N can be calculated by considering the values obtained for the scalar (dot) product of allowed reflections $[hkl]^*$ with electron beam directions $[uvw]$. For a f.c.c. crystal the value of N for the first zone is 1, if $(u + v + w)$ is odd and 2 if $(u + v + w)$ is even. For a b.c.c. crystal $N = 2$ if u, v, and w are all odd and $N = 1$ for all other values of u, v and w.

INTERPRETATION OF DIFFRACTION INFORMATION 87

4.10 are immediately obvious as seen from the equivalent circle drawn in Fig. 4.11(d).

Note that the measured radius of the first order zone on Fig. 4.10 is about 3.5 Å^{-1} (the size of any of the indexed zero order **g**'s can be used to calibrate distances) and, using equation (4.8), the calculated radius is found to be 3.53 Å^{-1}.

As mentioned earlier the fact that high order zones are being excited is demonstrated by fine lines visible in the direct beam in a convergent beam diffraction pattern (see Fig. 4.12). These lines are the high order zone equivalents of Kikuchi lines. Thus they can be indexed by pairing the dark lines through the zero order disc with the bright lines passing through the maxima in the FOLZ; the indices marked on some of the lines in Fig. 4.12 have been arrived at in this way. Alternatively, and equivalently, it is possible to construct patterns which are the high order zone equivalents of Kikuchi maps (see Appendix C), and hence index the lines by inspection.

Because these lines are very sharp and because they correspond to high angle reflections they are very sensitive to small changes in interplanar

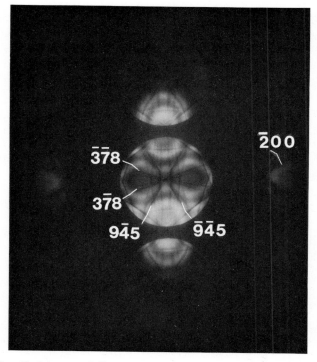

Fig. 4.12 Magnified image of direct beam from Fig. 4.11. The indexed lines are the HOLZ lines corresponding to some of the maxima on the first Laue zone (see text).

Table 4.3 Table showing the relation between the observed symmetries in convergent beam diffraction patterns and the 31 diffraction groups which correspond to the 32 different three-dimensional point groups. See text for full discussion.

Observed symmetry in zero order zone	Projection diffraction group	Possible diffraction groups (see Appendix B)	Symmetries of high order information	
			whole pattern	zero order disc
1	1_R	1	1	1
		1_R	1	2
2	21_R	2	2	2
		2_R	1	1
		21_R	2	2
m	$m1_R$	m_R	1	m
		m	m	m
		$m1_R$	m	2m
2mm	$2mm1_R$	$2m_R m_R$	2	2mm
		2mm	2mm	2mm
		$2_R mm_R$	m	m
		$2mm1_R$	2mm	2mm
4	41_R	4	4	4
		4_R	2	4
		41_R	4	4
4mm	$4mm1_R$	$4m_R m_R$	4	4mm
		4mm	4mm	4mm
		$4_R mm_R$	2mm	4mm
		$4mm1_R$	4mm	4mm

	31_R			$3m1_R$			61_R			$6mm1_R$		
3	3	3	3									
	3_R	3	6									
$3m$				$3m_R$	3	$3m$						
				$3m$	$3m$	$3m$						
				$3m1_R$	$3m$	$6mm$						
6							6	6	6			
							6_R	3	3			
							61_R	6	6			
$6mm$										$6m_Rm_R$	6	$6mm$
										$6mm$	$6mm$	$6mm$
										6_Rmm_R	$3m$	$3m$
										$6mm1_R$	$6mm$	$6mm$

spacing and provide, in principle, an accuracy of a few parts in 10^4 for lattice parameter determination. Unfortunately energy losses suffered by the electrons will give rise to broader lines, although energy filtering (see Chapter 3) can be used to remove this problem.

(c) Analysis of convergent beam diffraction patterns to yield point groups and space groups

In Section 4.2.1(c) it was shown that the crystal system of a specimen could be determined from a series of conventional electron diffraction patterns. Taken in conjunction with EDX and/or EELS analysis and measurement of the lattice parameter these observations will sometimes allow a complete identification of a specimen. However, if complex phases, which contain many atoms in a unit cell, are to be considered, this technique will not in general lead to a solution. It is because of this that CBDPs are so important because they do allow point groups and space groups to be determined and these, taken in conjunction with EDX/EELS, will virtually always enable a phase to be identified.

The simplest way to understand how to analyse CBDPs is to consider an example. The explanation requires that the reader be familiar with the concept of symmetry operations, point groups, diffraction groups and space groups. These subjects are covered in adequate detail in the references [5], [6] and [7]. Diffraction group symmetries are shown in Appendix B.

There are basically four stages involved in the complete determination of a crystal structure from CBDPs. These are:

(1) Determination of the projection diffraction groups and hence the diffraction group and the point group to which the crystal belongs.
(2) Determination of the type of unit cell and hence indexing of diffraction maxima. (As pointed out earlier this can be done using conventional diffraction patterns but when small precipitates are being analysed it is necessary to use a small, and therefore convergent probe. Hence this determination is considered part of the analysis using CBDP.)
(3) Determination of space group.
(4) Identification of the material. This requires chemical information (obtainable from EDX and EELS) which, taken with the conditions governing the intensities of the observed reflections and with the unit cell size, should allow unique identification to be made. In the case of a new material it will not always be possible to define atom positions, but possible positions will be defined in the appropriate space group description in [7].

These four points will now be dealt with in turn.

(i) *Determination of the projection diffraction group, diffraction group and hence the point group*

Because the symmetry of the whole convergent beam pattern (i.e. including the high order zones) must correspond to one of the ten two-dimensional point groups [5] analysis of the symmetry of the whole pattern enables possible point groups to be assigned. The symmetry of the zero order Laue zone may be used to determine the projection diffraction group and subsequent examination of the symmetry of the three-dimensional information in the pattern allows the diffraction group within the specific projection diffraction group to be assigned. This leads to the possible point group as discussed below.

The first assessment to be made from a CBDP is thus the symmetry of the zero order Laue zone, in order to determine the projection diffraction group to which the crystal belongs. This assessment requires that high order Laue zone information be excluded* when determining the symmetry of the zero order layer. The possible symmetries which can be observed are listed in the first column of Table 4.3. These symmetries are simply those of the ten two-dimensional point groups, i.e. the projection diffraction groups with the inversion operation removed. Each projection diffraction group contains several possible diffraction groups which are listed in the third column and the next stage in the analysis is to use the high order Laue information to determine the diffraction group. The possible symmetries shown by both the whole pattern (including high order information) and by the HOLZ lines in the zero order beam are shown in the next two columns for each of the possible diffraction groups. These symmetries, taken together with the knowledge of the projection diffraction group, enable possible diffraction groups to be identified. An important point to note is that the identification of the projection diffraction group from the zero layer pattern reduces the ambiguities present. For example, in columns four and five of the table, m symmetry is observed both for the whole pattern (including HOLZ) and for the HOLZ lines information in the zero order disc in diffraction groups m and $2_R mm_R$. These are, however, in different projection diffraction groups and are thus distinguished by using the zero layer information. In one case the zero layer would give m symmetry (corresponding to a projection diffraction group of $m1_R$ and in the case under consideration to a diffraction group of m) and the other would give $2mm$ symmetry (corresponding to a projection diffraction group of $2mm1_R$, and in the case under consideration to a diffraction group of $2_R mm_R$).

On the above basis possible diffraction groups are obtained and are related to point groups using Table 4.4. If CBDPs are taken down several

---

* The information within the discs is sensitive to small changes in thickness whereas the HOLZ are not. Examination of CBDPs taken at slightly different thicknesses may help in distinguishing zero order information from HOLZ lines.

Table 4.4 Relation between the diffraction groups and crystal point groups (after reference 5).

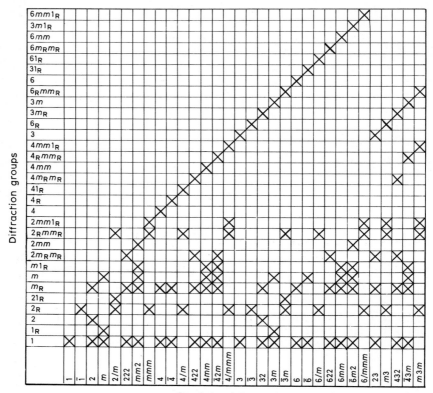

zone axes and the corresponding diffraction groups obtained as above then the point group can usually be determined unambiguously. If ambiguities still exist these can be removed by taking further CBDPs. Before dealing with how these additional CBDPs are selected systematically, it is useful to consider the derivation of the space group of a specimen from a series of CBDPs.

Fig. 4.13 shows two CBDPs taken in two different directions from the same sample and these will be used in the following discussion to illustrate the determination of space groups. Thus examination of Fig. 4.13(a) shows that the zero layer symmetry is described by $2mm$ and it follows (Table 4.3) therefore that the projection diffraction group is $2mm1_R$. Examination of the *whole* pattern (i.e. including not only the HOLZ lines through the bright field disc but the high order Laue zones themselves) shows that $2mm$ symmetry is again clearly shown. This whole pattern analysis therefore shows (Table 4.3) that the diffraction group is either $2mm1_R$, or $2mm$. A similar examination

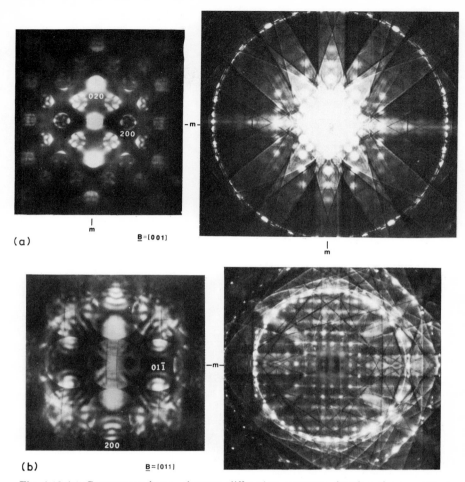

Fig. 4.13 (a) Convergent beam electron diffraction patterns showing the symmetry information in the zero order layer and in the high order Laue zone. For discussion see text. (b) Convergent beam electron diffraction patterns taken from the same sample as the patterns shown in (a). Patterns show zero layer symmetry information and symmetry information from high order zones. For discussion see text (after reference 23).

of Fig. 4.13(b) is summarized together with the anlysis of Fig. 4.13(a) in Table 4.5. As is clear from this table, the determination of the point group of the sample is incomplete, since the observed symmetries in the two different beam directions used for Fig. 4.13(a) and (b) have mmm, $4/mmm$, $6/mmm$, $m3$ and $m3m$ in common. Further information is required in order to resolve this problem and this is obtained both from other CBDPs and from the angle of tilt between these patterns (cf. Fig. 4.7).

Table 4.5 Table showing the observed symmetries of the zero order Laue pattern in Fig. 4.13(a) and (b) and the corresponding projection diffraction groups, and hence diffraction groups, which could yield the observed symmetries. The table also shows the observed symmetries of the whole diffraction pattern and the corresponding diffraction groups within the possible groups defined by the zero order symmetry. Finally the possible point groups, deduced from Table 4.4 are shown in the last column.

Figure number	Observed zero layer symmetry	Deduced projection diffraction group (see Table 4.3)	Possible diffraction groups (see Table 4.3)	Observed symmetry of whole patterns	Possible diffraction groups (see Table 4.3) within projection diffraction group	Possible point groups (Table 4.4)
4.13(a)	$2mm$	$2mm1_R$	$2m_Rm_R$ $2mm$ 2_Rmm_R $2mm1_R$	$2mm$	$2mm$ $2mm1_R$	$mm2$ $6m2$ mmm $4/mmm, 6/mmm$ $m3$ $m3m$
4.13(b)	$2mm$	$2mm1_R$	$2m_Rm_R$ $2mm$ 2_Rmm_R $2mm1_R$	m	2_Rmm_R	$2/m$ mmm $4/m, 4/mmm$ $\bar{3}m, 6/m, 6/mmm$ $m3, m3m$

It is very useful to have the basic symmetries of the crystal systems and the CBDP symmetries expected for various electron beam direction available when considering how to remove ambiguities of the sort shown in Table 4.5.

In the case under consideration the point group could be shown to be *mmm* since, firstly, no patterns were observed which showed fourfold or sixfold symmetries expected for the other possible point groups (see Table 4.6) and secondly, it was found that the three beam directions which showed $2mm1_R$ symmetry were mutually orthogonal.

Thus, at this stage the sample is known to be orthorhombic and since the three $\langle 100 \rangle$ directions have been identified the three lattice constants (or multiples of them) are available. The next step then is to attempt to determine the type of unit cell.

(ii) *Determination of type of unit cell*
If the unit cell is primitive, face centred or body centred (P, C, F or I) the CBDPs taken along $\langle 001 \rangle$ can sometimes be used to determine the type of unit cell*. This is done by projecting the reflections in the first and second order Laue zones onto the zero order layer in an $\langle 001 \rangle$ diffraction pattern. If a primitive cell is obtained then the real crystal is primitive; if a cell is obtained which is recognizably all-face centred then the reciprocal lattice is face centred and the real lattice body centred and vice versa. For base or side centred cells this technique will not be unambiguous since the projection will be dependent on the particular $\langle 001 \rangle$ along which the projection is made. Additionally this assessment can be misleading if there is a significant number of forbidden reflections.

The main aim behind the attempt to determine the type of unit cell is to limit the number of possible space groups which need be considered. The next step is to determine the possible space group to which the specimen belongs.

(iii) *Determination of space group*
In order to determine the space group it is necessary to identify forbidden reflections which occur due to double diffraction (see Appendix A). In CBDPs some forbidden reflections reveal dynamic absences [8], [9] which take the form of dark lines within the forbidden reflection. The fact that they are forbidden reflections can be confirmed by tilting about the systematic row which contains the reflection. If they are indeed due to double diffraction they will disappear (see Appendix A).

The presence of these forbidden reflections implies either that there is a glide plane which contains both **B** and the forbidden reflection and/or that there is a twofold screw axis perpendicular to **B**. In order to ascertain whether both or one of these is present, the crystal is tilted about g_f and

---
* Note also the conditions for allowed reflections in Appendix A.

Table 4.6 Table showing the symmetries which will be observed in convergent beam diffraction patterns from crystals of the 32 point groups in the electron beam directions indicated (after reference 5).

Crystal system	Point group	⟨111⟩	⟨100⟩	⟨110⟩	⟨uv0⟩	⟨uuw⟩	⟨uvw⟩
Cubic	$m3m$	$6_R mm1_R$	$4mm1_R$	$2mm1_R$	$2_R mm_R$	$2_R mm_R$	2_R
	$\bar{4}3m$	$3m$	$4_R mm_R$	$m1_R$	m_R	m	1
	432	$3m_R$	$4m_R m_R$	$2m_R m_R$	m_R	m_R	1
	$m3$	6_R	$2mm1_R$		$2_R mm_R$		2_R
	23	3	$2m_R m_R$		m_R		1

		[0001]	⟨11$\bar{2}$0⟩	⟨1$\bar{1}$00⟩	⟨uvt0⟩	⟨uutw⟩	⟨u$\bar{u}$0w⟩	⟨uvtw⟩
Hexagonal	$6/mmm$	$6mm1_R$	$2mm1_R$	$2mm1_R$	$2_R mm_R$	$2_R mm_R$	$2_R mm_R$	2_R
	$\bar{6}m2$	$3m1_R$	$m1_R$	$2mm$	m	m_R	m	1
	$6mm$	$6mm$	$m1_R$	$m1_R$	m_R	m	m	1
	622	$6m_R m_R$	$2m_R m_R$	$2m_R m_R$	m_R	m_R	m_R	1
	$6/m$	61_R			$2_R mm_R$			2_R
	$\bar{6}$	31_R			m			1
	6	6			m_R			1

		[0001]	⟨11$\bar{2}$0⟩	⟨1$\bar{1}$00⟩	⟨uvt0⟩	⟨uutw⟩	⟨u$\bar{u}$0w⟩	⟨uvtw⟩
Trigonal	$\bar{3}m$	$6_R mm_R$	21_R				$2_R mm_R$	2_R
	$3m$	$3m$	1_R				m	1
	32	$3m_R$	2				m_R	1
	$\bar{3}$	6_R					2_R	2_R
	3	3					1	1

		[001]	⟨100⟩	⟨110⟩	[u0w]	[uv0]	[uuw]	⟨uvw⟩
Tetragonal	$4/mmm$	$4mm1_R$	$2mm1_R$	$2mm1_R$	2_Rmm_R	2_Rmm_R	2_Rmm_R	2_R
	$\bar{4}2m$	4_Rmm_R	$2m_Rm_R$	$m1_R$	m_R	m_R	m	1
	$4mm$	$4mm$	$m1_R$	$m1_R$	m	m_R	m_R	1
	422	$4m_Rm_R$	$2m_Rm_R$	$2m_Rm_R$		2_Rmm_R		1
	$4/m$	41_R				m_R		2_R
	$\bar{4}$	4_R				m_R		1
	4	4						1
Orthorhombic		[001]	⟨100⟩	[u0w]	⟨110⟩	[uv0]	[uvw]	
	mmm	$2mm1_R$	$2mm1_R$		2_Rmm_R	2_Rmm_R		
	$mm2$	$2mm$	$m1_R$		m	m_R		
	222	$2m_Rm_R$	$2m_Rm_R$		m_R	m_R		
Monoclinic		[010]	[u0w]	[uvw]				
	$2/m$	21_R	2_Rmm_R	2_R				
	m	1_R	m	1				
	2	2	m_R	1				
Triclinic		⟨uvw⟩						
	$\bar{1}$	2_R						
	1	1						

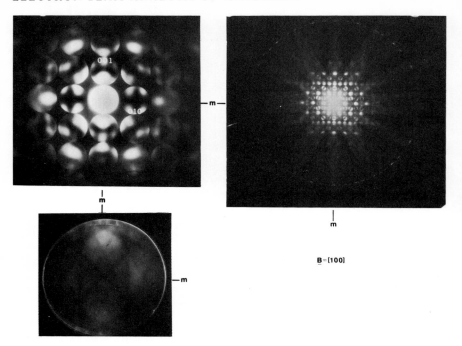

Fig. 4.14 Convergent beam diffraction pattern taken from the same sample as in Fig. 4.13. The 010 reflections on this pattern are marked and further tilting shows that reflections are due to double diffraction. See text for discussion (after reference 23).

perpendicular to g_f, where g_f is the forbidden reflection. If the crystal is tilted perpendicular to g_f, and on continued tilting another forbidden reflection is observed, then this observation confirms that a glide plane is present. If the crystal is tilted about g_f (so that it disappears) and g_f reappears at a different **B**, then a twofold screw is present.

Dynamic absences can be seen in the [100] (see Fig. 4.14) and also in the [101] and [110] CBDPs. Since the dynamic absence occurs for $g_f = \pm 010$, in $\mathbf{B} = [100]$ and it reappears in $\mathbf{B} = [101]$ there is a twofold screw axis parallel to the tilt axis [010]. The occurrence of different dynamic absences (i.e. of forbidden reflections) in both [010] and [110] shows that there is a glide plane parallel to (001). The general rule for kinematically absent reflections observed in all the diffraction patterns was found to be $h00$, $0k0$, h or $k = 2n + 1$; $hk0$, $h + k = 2n + 1$.

With this information, i.e. that the cell is orthorhombic, the point group is *mmm*, the lattice is primitive and there is both a twofold screw axis and a glide plane together with the rule for kinematically absent reflections the space group [7] is defined as P*mmn*. It remains then to identify the chemical formula and the atomic positions of the atoms in the crystal.

(iv) *Identification of specimen*

If we consider the example discussed above then the additional information available from EDX shows that it is a NiMo alloy with an atomic ratio corresponding to the formula Ni_3Mo. Measurements made on diffraction patterns at each of the poles showed that the lattice parameters were in close agreement with those of Ni_3Mo. The forbidden reflections listed under P*mmn* correspond with those obtained for the appropriate unit cell and for this structure the atom positions are listed in [7]. Thus the analysis is complete and serves to illustrate the power and value of CBDP analysis.

The above account may appear to be very complex because all possibilities were considered. In practice the determination of the point group is fairly rapid if sufficient CBDPs have been taken. The determination of the space group is more time consuming since the behaviour of forbidden reflections on tilting is normally investigated as a specific exercise after the first series of patterns have revealed their presence.

It is worth summarizing the procedure so that the various steps are clearly defined.

(v) *Summary of procedure for obtaining and analysing convergent beam diffraction patterns*

(1) Take a series of CBDPs at each pole. Initially it is sensible to record patterns for at least three different, low order beam directions (cf. Section 4.2.1(c)). Additional patterns may be required and this will become apparent if ambiguities still exist, as discussed below. Use both a large and a small camera constant so that high angle and detailed information within discs can be observed. Choose the size of the second condenser aperture such that the discs just touch for the smallest reciprocal spacing. Tolerate a small overlap rather than make discs too small. Ensure that the beam is precisely along the zone axes, making the final, very small ($< 0.1°$), adjustments using the position of the C2 aperture. Note tilt settings at each pattern. At each orientation also take a pattern using a very small C2 aperture, or take a selected-area pattern if appropriate, and use these patterns for measuring parameters and for measuring angles between diffracting vectors. Focussing of the patterns with the diffraction focus control must be varied to suit each specific requirement. Thus when recording HOLZ lines through the zero order disc the sharpness of the image of the C2 aperture can be used as an aid to focussing if the HOLZ are difficult to see. If specific attention is being paid to the outer high order circles of diffraction maxima it is preferable to focus to minimize the distortion of these rings.

Note that these patterns, taken at three different electron beam directions will allow an identification of the crystal system (cf. Section 4.2.1(c)).

(2) Use the symmetries of the zero order layer to determine the projection diffraction group.
(3) Use the symmetries in the high order zones (whole pattern and bright field disc) to determine possible diffraction groups within the projection diffraction group.
(4) Use the possible diffraction groups to identify the three-dimensional point groups which will confirm the crystal system. If the point group cannot be identified unambiguously additional CBDPs must be taken and selection of these required patterns can be done with the aid of Tables 4.4 and 4.6.
(5) With the crystal system known, index the patterns and project the high order diffraction information in an [001] pattern onto the zero order layer to determine if possible whether the lattice is P, F or I. Note indices of reflections which are present which will also help distinguish between P, F or I (see Appendix A).
(6) Identify the kinematically forbidden reflections which may appear in some diffraction patterns and tilt both parallel and at an angle to these reflections to assess whether there is a twofold screw axis and/or a glide plane.
(7) Refer to the table of space groups in [7] to define the space group.
(8) If the chemical composition of the sample is known from EDX or EELS then this information, together with the space group, commonly allows identification of the phase. (A reference book (e.g. [10]) which lists space groups for many compounds, is very useful.) If the phase is not a known phase it is generally not possible to define atomic positions and if a further description is required it would be necessary to know for example, the density of the phase. This is generally not possible with the type of work considered in this book since it is envisaged that we are examining small precipitates which are transparent to 100 kV electrons.

One point should be made which concerns the general applicability of the CBDP method. If CBDPs are to be obtained which exhibit the full symmetry of the crystal it is essential that the region examined be fault free. It is further essential that the dislocation and point defect density be low and for HOLZ observations the effective temperature should be low so that HOLZ are observed.

Finally it should be noted that additional symmetry information is contained in convergent beam patterns if use is made of symmetries of diffracted beams, but as is clear from the above discussion, this information is not necessary for solving CBDPs. These other symmetries are discussed in [5].

4.3 INTERPRETATION OF DIFFRACTION MAXIMA ASSOCIATED WITH PHASE TRANSFORMATIONS AND MAGNETIC SAMPLES

Transmission electron diffraction patterns can contain regions of either sharp or diffuse intensity between the main diffraction maxima. In most cases the qualitative interpretation of these regions of intensity is reasonably straightforward, but in all cases quantitative interpretation is difficult. The origin of the difficulties lies, in the main, with the strong scattering of electrons by matter so that multiple scattering must always be important and can virtually never be accurately calculable. In the following sections the various factors which can give rise to these diffraction maxima will be briefly discussed.

4.3.1 Long range order

The simplest case to consider is that of an alloy which possesses long range order. The relative kinematic intensities of fundamental (I_F) and superlattice (I_s) intensities are related to the degree of order S by the well known relation

$$S = \left(\frac{I_s}{I_F}\right)^{1/2} \tag{4.12}$$

The kinematic intensities are determined by the structure factors for the various reflections and considering for example NiAl with $S = 1$ the expression for F_{hkl} can be written as

$$F_{hkl} = f(\text{Ni}) \exp[-2\pi i(hu_{\text{Ni}} + kv_{\text{Ni}} + lw_{\text{Ni}})] \\ + f(\text{Al}) \exp[-2\pi i(hu_{\text{Al}} + kv_{\text{Al}} + lw_{\text{Al}})] \tag{4.13}$$

where u_{Ni}, v_{Ni}, w_{Ni} are the coordinates of the Ni atoms, $u_{\text{Al}}v_{\text{Al}}w_{\text{Al}}$ are the coordinates of the Al atoms and hkl are the indices of the reflection. If we take all the Ni atoms to be at 0, 0, 0 and all the Al atoms at $\frac{1}{2}, \frac{1}{2}, \frac{1}{2}$ then it can be easily shown (using the facts that $e^0 = 1$, $e^{2\pi i} = 1$ and $e^{\pi i} = -1$) that when $h + k + l$ is odd

$$F = f(\text{Ni}) - f(\text{Al}) \tag{4.14}$$

and when $h + k + l$ is even

$$F = f(\text{Ni}) + f(\text{Al}) \tag{4.15}$$

In principle the degree of order can be obtained from the measured*

---

* Microdensitometry of negatives is the standard way of measuring intensities but postspecimen scanning and STEM allow on-line intensity measurements to be carried out.

intensities because, if all the Ni atoms are not at 0, 0, 0 or equivalent positions, the intensity of maxima for which $h + k + l$ is odd will be correspondingly reduced (and similarly for the Al atoms) because the intensity of these reflections is determined by the difference in the scattering factors of the atoms occupying the respective sites. As discussed in Section 4.2 the degree of long range order can be obtained only if the foil thickness and the electron beam direction are carefully measured, and if the necessary calibration of intensities has been carried out. The problems in interpretation of these diffraction maxima arise because multiple scattering between beams changes relative intensities, and the degree of multiple scattering is strongly influenced by the precise electron beam direction and by the specimen thickness.

If a specimen consists of ordered precipitates within a matrix the diffraction pattern will be made up of the two individual patterns from precipitate and matrix together with maxima generated by double diffraction (see Appendix A). This can result in rather complex patterns, but the significance of double diffraction in generating extra diffraction maxima can be assessed by tilting about various defined axes (cf. Section 4.2 and Appendix A).

Thus the presence of long range order may manifest itself by the presence of superlattice reflections, the intensities of which can, with care, allow S to be determined. As discussed in Chapter 5, antiphase domains can be imaged if dark field images are obtained by imaging with appropriate superlattice reflections. This is a valuable aid in confirming that the maxima are due to the presence of long range order.

Electron diffraction patterns can reveal the presence of long range order even when the X-ray scattering factors of the two elements are very similar. For example, superlattice reflections are easily seen in electron diffraction patterns of β-brass (Cu–Zn), but it is by no means easy to see the corresponding maxima in X-ray patterns. The superior peak to background ratio on electron diffraction patterns and the contribution of scattering from the nucleus, underly the improved visibility of the superlattice relections in Cu–Zn and similar alloys.

4.3.2 Short range order

The presence of short range order in a sample leads to regions of diffuse intensity, rather than sharp diffraction maxima between the fundamental reflections. Qualitative interpretation of diffuse scattering caused by short range order shows that the regions of diffuse intensity will be associated with those parts of a diffraction pattern where a superlattice reflection would appear – as is clear if short range order is due to small regions of order within a disordered matrix. As the degree of order within the small ordered regions decreases so the diffuse scattering will get weaker. The interpretation of the diffuse scattering is made easier by varying the heat treatment to see if the scattering behaves as would be expected for an order–disorder transformation. Difficulties arise when the degree of short range order is required from the intensity of the diffuse scattering.

The quantitative interpretation of diffuse X-ray scattering, caused by short range order, was developed about thirty years ago (e.g. [11], [12]), but because the theory is a kinematic theory it is not generally applicable to diffuse scattering of electrons. Quantitative analysis of diffuse scattering firstly requires the subtraction of the background contribution associated with other scattering phenomena. Thus, inelastic scattering including Kikuchi lines, and other diffuse scattering mechanisms, such as scattering caused by static and thermal displacements of atoms, will all contribute to the diffuse background. For realistically thick samples there will be multiple scattering so that the intensity of each of these factors will be influenced by the intensity of all the others. There is no theory which allows accurate subtraction of these contributions, and the best that can be done is to choose experimental conditions which minimize these complexities (avoiding high symmetry directions, use thin samples, tilt to avoid Kikuchi lines being near the region of interest, etc.) and then to assume that the background is a smoothly varying function of angle. The diffuse scattered intensity associated with specific regions of reciprocal space can, on this basis, be obtained and interpretation then requires that models of the sample be assumed and the diffuse intensity appropriate to the model be calculated and compared with the observation. These calculations are generally kinematic because the assumption of a smoothly varying non-dynamical background make dynamical calculations inappropriate. The equations relating calculated intensity to the short range order parameters are given in the X-ray papers [11], [12], [13]. At present it would appear that other diffraction techniques are more suited to assessing short range order.

4.3.3 Precipitation and pretransformation diffuse scattering

Solid solutions can give rise either to extra reflections or to diffuse scattering associated with several different precipitation and transformation phenomena. Thus a solid solution can give rise to a two-phase mixture consisting of a solid solution and an ordered phase which may occur via a spinodal mechanism [13]. Ordered precipitates can be nucleated initially at random (as discussed above for the case of short range ordered alloys) which then align as they grow along specific soft directions in the crystals. Additionally, continuous ordering can occur in which the degree of short range order appears to increase continuously. Combinations of these reactions are possible as well as premartensitic phenomena [14] and ω-transformations. Identification and interpretation of the extra diffraction information can be complex. In all cases it is important to correlate image and diffraction information and to assess changes caused by different heat treatments.

(a) Spinodal decomposition

Spinodal decomposition involves the formation of solute-rich and solute-lean regions in the sample, with ideally only one wavelength (typically ~ 10 nm) defining the distance between solute-rich (lean) regions, but in an imperfectly formed spinodal there is commonly a range of wavelengths. The existence of such modulations in the alloy along a specific direction in a crystal leads to diffraction maxima formed in that direction, the separation of which defines the (reciprocal) of the wavelength; when a range of wavelengths is present these maxima will be diffuse. These extra reflections appear as satellites associated with the main diffraction maxima because the spinodal wavelength is so much larger ($\sim$ fifty times) than the interplanar spacing. The periodicity will also be revealed in the modulations of interplanar spacings visible in the image since the objective aperture will accept the beam due to the modulation (**g** typically ~ 0.01 Å^{-1}) together with the direct (b.f.) or diffracted (d.f.) beam so that they will interfere in the image plane (cf. Chapter 5).

Confirmation that satellites are due to a spinodal reaction is done by carrying out heat treatments which would be expected to change the perfection of the spinodal structure and assessing the corresponding changes in image and diffraction space. It should be emphasized that the long spacings associated with spinodals, which are typically about fifty times the interatomic spacings, have the consequence that the satellite reflections can easily be lost in the much stronger fundamental reflections. Defocussed illumination and long exposures are essential for successful examination of spinodal diffraction data. Post-specimen scanning of a defocussed SAD pattern across a STEM detector, and subsequent display of the Y-modulated pattern, is a good way to examine spinodal patterns.

(b) Cluster formation

The formation of small clusters at random commonly results in diffuse scattering (short range order is an example of this) which is clearly visible in diffraction patterns. An example taken from Ni-rich NiAl is shown in Fig. 4.15 where the two diffraction patterns have been taken from the same area of the sample after tilting to the two different matrix electron beam directions indicated. It can be seen that the visible details of the diffuse scattering change as the sample is tilted, and by tilting about several axes it is possible to build up a picture of the three-dimensional surface which gives rise to the diffuse scattering in the diffraction pattern in the various two-dimensional sections of reciprocal space visible in each pattern. This surface passes through the positions where maxima will form from the

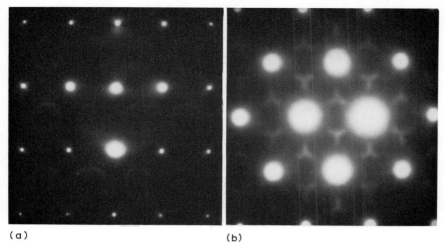

Fig. 4.15 Transmission electron diffraction patterns taken from a sample of Ni 48% Al showing diffuse scattering between the main diffraction maxima: (a) $\mathbf{B} = [110]$; (b) $\mathbf{B} = [111]$.

ordered phase or from the new phase; in the case shown in Fig. 4.15 this new phase is Ni_2Al_3. Examination of the periodicity of the diffuse scattering shows for example in the present case, that it repeats every $[100]^*$ rather than $[200]^*$, showing that the cluster responsible for the scattering lies of one or other of the sublattices and the maxima appear to lie on a distorted in-sphere of the $\{111\}^*$ reciprocal lattice planes.

Thus the qualitative interpretation of the diffuse maxima caused by clusters is carried out by means of tilting experiments so that the three-dimensional surface is defined and the composition and/or heat treatments are changed so that the nature of new phases or precipitates which then form may be investigated. Quantitative interpretation requires, as was discussed for short-range order, estimation of the experimental intensities and their comparison with calculations of the intensities expected from modelled clusters.

(c) Premartensitic phenomena

Alloys such as NiTi which undergo a martensitic reaction show complex diffraction patterns if they are examined at temperatures just above the transformation temperatures (e.g. [15]). These maxima have been interpreted in terms of charge density waves (CDW) and, because several such waves can be present simultaneously, the modelling of the observations is difficult. Effectively these CDW are associated with regions of the sample over which there are periodic displacements, and, following wave terminology, these displacements can be described by a propagation wavevector and

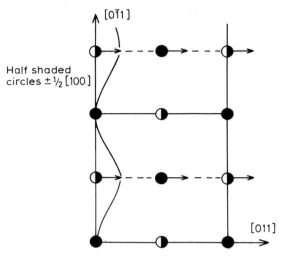

Fig. 4.16 Schematic diagram illustrating the role of a charge density wave (CDW) in displacing atoms in a crystal. These waves are present in alloys just above the martensite transformation temperature. The diagram shows a [100] projection of NiAl with a $\frac{1}{2}[0\bar{1}1][011]$ transverse wave causing the displacements indicated.

by a polarization vector **e**. Fig. 4.16 shows schematically the way in which a wave within a crystal can describe the displacements which may be involved in a shear transformation.

The interpretation of the complex diffraction data is best done in conjunction with dark field imaging and HREM. These techniques are capable of showing immediately the scale of the regions that are responsible for the diffraction maxima. Measurements of fringe spacing on HREM micrographs can be used to confirm the commensurate or incommensurate nature of the waves (the waves are termed commensurate if they have a wavelength which is a simple multiple of the crystal spacings). The interpretation of these premartensitic phenomena, and their precise relation to the actual martensite transformation are the subject of considerable theoretical and experimental work at present and it is not possible to give a definitive account of the characteristics expected in diffraction patterns. Because weak reflections are to be expected, some care is required if they are to be detected, as was discussed earlier. Nevertheless if CDWs are modelled then the influence on diffraction patterns can be calculated.

Thus if the atoms of the CDW are at u_i, v_i, w_i, the structure factor for the extra reflection will be given by

$$F = \sum_i f_i \exp\{-[2\pi i(hu_i + kv_i + lw_i)]\} \tag{4.17}$$

The kinematic intensities of these reflections will be given by

$$I \propto FF^* |\mathbf{g}\cdot\mathbf{e}|^2 \tag{4.18}$$

where F^* is the complex conjugate of F [16]. Hence by varying $\mathbf{g}$ the value of $\mathbf{e}$ can be determined using standard analysis, analogous to that used in defect analysis (see Chapter 5).

(d) ω-transformation

Extra reflections and diffuse scattering are also associated with transformation to the ω-phase. This transformation occurs in many b.c.c. alloys and is best understood by referring the b.c.c. cell to hexagonal axes [17]. On this basis the b.c.c. cell has atoms at $0,0,0$ and $2/3, 1/3, 1/3$ and $1/3, 2/3, 2/3$; the ω-phase has atoms at $0,0,0$ and $2/3, 1/3, (1/3 + u)$, and $1/3, 2/3, (2/3 - u)$ where u is ideally $1/6$ so that $(1/3 + u)$ and $(2/3 - u) = 1/2$. Therefore the ω-phase can give reflections defined by these new atomic positions whose kinematic amplitudes will be given by

$$A = f_i \sum_i \exp[-2\pi i (hu_i + kv_i + lw_i)] \tag{4.19}$$

where f_i is taken as the mean atomic scattering factor for the atoms in the b.c.c. alloy so that the amplitudes derived from equation (4.19) refer to disordered ω. (Clearly if ω is derived from an order alloy then superlattice ω reflections will be generated.) Equation (4.19) can be written

$$A = f_i \left\{ 1 + \exp\left[-2\pi i \left(\frac{2h}{3} + \frac{k}{3} + \frac{1}{2}\right)\right] + \exp\left[-2\pi i \left(\frac{h}{3} + \frac{2k}{3} + \frac{1}{2}\right)\right] \right\} \tag{4.20}$$

by substituting the appropriate atomic coordinates and hence the amplitudes of any relfection, hkl, calculated.

On this basis the presence of ω-phase in b.c.c. alloys can be detected fairly straightforwardly because sharp maxima would be expected of (kinematic) amplitude defined by equation (4.20). This, taken in conjunction with the changes in intensities of these maxima either on changing the temperature of the sample, or changing the composition, makes identification of ω-phase from diffraction patterns fairly straightforward. Again HREM is a very powerful aid in interpreting the diffraction data and in defining the spatial distribution of the ω.

In addition to the sharp ω maxima, diffuse scattering, displaced from the sharp ω reflections, is observed in some alloys but the interpretation of this is not straightforward. Various models involving for example the presence of a localized defect consisting of a vacancy and associated rows of atoms displaced along $\langle 111 \rangle$ have been put forward (e.g. [18], [19], [20]) but other techniques, such as neutron diffraction are better suited to the analysis.

4.3.4 Magnetic samples

Magnetic samples will give rise to extra spots in diffraction patterns because the electron beam will be deflected in different directions by the variously oriented domains. The angular displacement of the electrons from the incident beam direction is due to the Lorenz force (see Chapter 1 for the related discussion concerning the operation of magnetic lenses) which acts at right angles to the electron velocity. Typical deflection angles for 100 kV electrons are of the order of 5×10^{-4} radians [1] for iron, cobalt and nickel. Although this is only a small angle, split spots are visible on defocussed diffraction patterns and these can be used to image magnetic domains in several different ways (see Chapter 5).

4.4 INTERPRETATION OF DIFFRACTION PATTERNS FROM TWINNED CRYSTALS

It is evident that twinned crystals will generally give rise to more diffraction maxima than will a single crystal. In order to show that the extra maxima in a diffraction pattern arise from a twin it is necessary to show that the orientation relationship between the two patterns (from the two crystals) conforms to that expected for the particular twin law. The simplest case arises when the electron beam direction lies parallel to the composition plane since the twin pattern is generated from the matrix pattern simply by rotating about the twin axis – which must lie in the plane of the diffraction pattern for this geometry. For example, in the case of a twinned f.c.c. crystal with $\mathbf{B} = [101]$ with a twin on $(\bar{1}\bar{1}1)$ then the twin pattern will be generated by rotating the [101] pattern by 180° about the $\bar{1}\bar{1}1^*$ reflection. The composite patterns of twin and matrix will therefore have no extra diffraction maxima along the reciprocal lattice row through the origin which contains $\bar{1}\bar{1}1^*$, but because this row is not a mirror plane, there will be extra maxima in all other rows in the composite patterns. (A schematic [101] f.c.c. diffraction pattern is shown in Appendix C and can be used to show that rotation about $(\bar{1}\bar{1}1)$ will produce these extra maxima.) This type of diffraction pattern in which a central row contains fewer diffraction maxima than other parallel rows is characteristic of a twinned pattern with the twin axis perpendicular to $\mathbf{B}$ if the central row is not a mirror plane. The twin axis and twin plane are then defined from the indices of the diffraction pattern. In other orientations the interpretation will not be as straightforward but the following procedures may be used for any crystal system to assess whether two grains are twinned.

Because most electron microscopes have tilting facilities of at least $\pm 45°$ about two orthogonal axes it will commonly be possible to tilt the specimen so that the boundary, which will be visible in the image, is vertical. If the boundary is a coherent twin boundary the diffraction pattern will immediately define the crystallography since the diffracting vector perpendicular to the twin plane defines the twin axis. If the twin is incoherent or if the boundary cannot be tilted so that it is vertical, the above simple analysis cannot be used. In such cases in order to confirm the twin/matrix orientation relationship an analogous technique to that used in relation to Fig. 4.7 can be used. Thus diffraction patterns taken either side of the boundary, in at least two different electron beam directions for each grain, will define both the crystal system of the twin and the twin composition plane. In addition it is straightforward to plot the orientations of crystals on either side of a boundary on a composite stereogram. If the two crystals are twin-related this will be apparent.

Alternatively, it is obviously possible to calculate the positions of all extra diffraction maxima from a given twin law for any matrix beam direction and to compare the observed and calculated patterns. With complex crystals this requires many different orientations to be considered and, in view of the simplicity of the experimental approach discussed above, it seems preferable to carry out the tilting experiments rather than many calculations. Whichever technique is used, if the results are taken in conjunction with dark field imaging, unambiguous identification of twins can be carried out.

4.5 INTERPRETATION OF CHANNELLING PATTERNS AND BACKSCATTERED ELECTRON PATTERNS IN SCANNING ELECTRON MICROSCOPY

4.5.1 Channeling patterns

As discussed in Chapter 3 channelling patterns can be obtained by rocking the beam about a fixed point on the specimen surface. The backscattered electrons, or the specimen current, may be used to modulate the intensity on the CRT, and because the backscattering efficiency is a function of orientation, the patterns contain information which allows the orientation of the specimen to be obtained. When the electron beam is parallel to a set of planes there is a minimum in the backscattering intensity (i.e. electrons are channelled at this symmetry orientation) and the backscattering efficiency changes either side of this orientation [21]. This leads to patterns which are somewhat similar to Kikuchi patterns as shown in Fig. 4.17. When the mean electron beam direction is parallel to a prominent zone axis then a pattern is generated which reflects the symmetry of the crystal in this direction.

The larger the angle through which the beam is rocked the larger the angular range of the channelling pattern which is visible and the more easily is the pattern identified. However, the size of the area from which the channelling pattern is generated increases as the angle of rock increases.

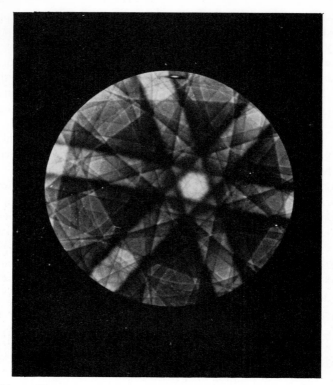

Fig. 4.17 Channelling pattern from a sample of silicon obtained using backscattered electrons in an SEM operating at 20 kV. The electron beam direction with respect to the sample is obtained most straightforwardly by comparing the symmetry with a channelling pattern map which is very similar to the Kikuchi maps shown in Appendix C.

Thus the non-axial rays at the extreme of the angle of rock will be focussed above the specimen plane and hence impinge on the specimen at a distance from the axial ray given by $C_s \alpha^3$, where C_s is the aberration coefficient of the probe-forming lens and α is the deviation of the ray from the optic axis. Typically C_s for a SEM is about 30 mm and the minimum area from which a channelling pattern covering 4° could be obtained is about 5 μm since $d_{min} = C_s \alpha^3 / 2$. It should be noted that for a STEM C_s is typically 2 mm and the minimum area for $\pm 4°$ channelling pattern would be about 0.3 μm. In addition it is clear, from the fact that the backscattering efficiency is a function of incident angle, that the patterns will show more contrast the smaller the beam divergence. This is a similar argument to that used when considering diffraction contrast imaging in STEM, since an averaging effect over a range of orientations is involved. The contrast in channelling patterns is low (typically $\sim 5\%$) and differentiation of the signal is commonly used to increase the visibility of the lines.

The simplest method of indexing channelling patterns is to compare individual patterns with a composite pattern which covers a sufficiently large angle. As with Kikuchi patterns it is possible, with experience, to index patterns on sight, and hence the angle from nearby poles is immediately known. Alternatively, if the pattern cannot be indexed the specimen can be tilted through a measured angle until an identifiable pole can be observed.

Fig. 4.18 Backscattered pattern obtained from a sample of silicon. The surface normal is clearly near [011] as can be seen by comparing the symmetry visible in this pattern with the Kikuchi maps in Appendix C.

4.5.2 Electron backscattering patterns

Electron backscattering patterns provide basically the same information as do channelling patterns but the angular resolution is better in the backscattering patterns [22]. Fig. 4.18 is an example of a backscattering pattern obtained from using the geometry discussed in Chapter 3. As discussed in Chapter 3 the electrons which make up the electron backscattering patterns are elastically scattered electrons which are subsequently Bragg diffracted by appropriately oriented planes. High angle elastic scattering provides a spherical source of electrons and the formation of these patterns is similar to that of Kikuchi lines.

Because the lines in backscattering are sharper than the lines in channelling patterns the accuracy of parameter measurements is higher, and because the patterns contain information covering such a large angle they are very useful for determining crystal symmetries.

REFERENCES

1. Hirsch, P.B., Howie, A., Nicholson, R.B., Pashley, D.W., and Whelan, M.J. (1965) *Electron Microscopy of Thin Crystals*, Butterworth, Sevenoaks.
2. James, R.W. (1950) *The Optical Principles of the Diffraction of X-rays*, Bell, London.
3. Thomas, L.E. and Humphreys, C.J. (1970) *Phys. Stat. Solidi*, **A3**, 594.
4. Kelly, P.M., Jostsons, A., Blake, R.G. and Napier, J.G. (1975) *Phys. Stat. Solidi*, **A31**, 771.
5. Buxton, B.F., Eades, J.A., Steeds, J.W. and Rackham, G.M. (1976) *Phil. Trans. R. Soc. A*, **281**, 171.
6. Kelly, A. and Groves, G.W. (1979) *Crystallography and Crystal Defects*, Longman, Harlow.
7. Henry, N F M and Lonsdale, K. (eds) (1952) *International Tables for X-ray Crystallography I: Symmetry Groups*, Kynoch Press, Birmingham.
8. Goodman, P. (1974) *Nature (London)*, **251**, 698.
9. Gjonnes, J. and Moodie, A.F. (1965) *Acta Cryst.*, **19**, 65.
10. Pearson, W.B. (1967) *A Handbook of Lattice Spacings of Metals and Alloys*, Vol. 2, Pergamon, Oxford.
11. Cowley, J.M. (1950) *J. Appl. Phys.*, **21**, 24.
12. Warren, B.E., Averbach, B.L. and Roberts, B.W. (1951) *J. Appl. Phys.*, **22**, 1493.
13. Laughlin, D.E. and Cahn, J.W. (1975) *Acta Met.*, **23**, 329.
14. Bagargatski, Iu. A., Nosova, G.I. and Tagunova, T.V. (1955) *Dokl. Akad. Nauk. SSSR*, **105**, 1225.
15. Sandrock, G.D., Perkins, A.J. and Hehemann, R.F. (1971) *Metall. Trans.*, **2**, 2769.
16. Guinier, A. (1963) *X-ray Diffraction in Crystals*, Freeman, San Francisco.
17. Silcock, J.M., Davies, M.H. and Hardy, K. (1956) *Symp. on the Mechanics of Phase Transformation in Metals*, Inst. Metall., London.
18. Cook, H.E. (1973) *Acta Met.*, **21**, 1445.
19. Chang, A.L.J., Sass, S.L. and Krakov W. (1976) *Acta Met.*, **24**, 29.
20. Kuan, T.S. and Sass, S.L. (1976) *Acta Met.*, **24**, 1053.
21. Booker, G.R., Shaw, A.M.B., Whelan, M.J. and Hirsch, P.B. (1967) *Phil. Mag.*, **16**, 1185.
22. Dingley, D.J. (1983) *Scanning Electron Microscopy*, (ed. O. Johari), Chicago Press, Chicago.
23. Kaufmann, M.J., Eades, J.A., Loretto, M.H. and Fraser, H.L. (1983) *Metall. Trans. A*, **14A**, 1561.

5
ANALYSIS OF MICROGRAPHS IN TEM, STEM, HREM AND SEM

5.1 INTRODUCTION

The images obtained in electron microscopes are influenced to greater or lesser extents by the diffraction conditions. In TEM and STEM the diffraction conditions constitute the most important factors in influencing the contrast, whereas in SEM these factors are generally not important. As discussed in Chapter 3 it is possible to obtain diffraction patterns from very small regions in TEM and STEM and if a diffraction pattern is obtained from the area from which the micrograph is obtained then the micrograph can be interpreted. In this section it will be assumed that the diffraction information listed in Section 4.2.1 (the electron beam direction **B**, the indices of the planes giving rise to the image **g**, and the deviation from the Bragg condition s_g) has been obtained and is therefore available for interpreting the corresponding micrograph.

In Chapter 3 it was pointed out that an objective aperture is used in TEM to select either the direct or diffracted beam to form the image and in STEM a detector is used to select the imaging electrons. Exactly equivalent images are obtained in TEM and STEM if the incident convergence angle in TEM is the same as the collection angle in STEM and vice versa [1]. In the following sections the interpretation of images will be discussed without differentiating between TEM and STEM images, since in principle identical images can be obtained in transmission using the two techniques. In practice very different electron optical conditions are invariably used and this leads to understandable differences in the images. The significance and origin of these differences are discussed later in this section.

If the direct beam is used to form the image a bright field image is obtained and if a diffracted beam is used a dark field image is formed. The electron images are simply magnified images of the electron intensity on the bottom surface of the specimen, and contrast arises only if the intensity varies significantly from one region of a specimen to another. The contrast observed will then be a function of the diffraction conditions; planes near to the Bragg condition will diffract strongly if they are locally bent to the Bragg condition.

In this section a simple description of the theories which have been developed to interpret electron micrographs will be given and it will be seen that these theories will require that **B**, **g** and s_g be known.

Two simplifying assumptions are usually made in these theories. Firstly it is assumed that the intensity on the bottom surface can be obtained by calculating independently the intensity at the bottom of columns about 2.0 nm in diameter. Secondly, it is assumed that only one set of planes is at or near the Bragg condition, i.e. it is assumed that only two electron beams need be considered: the direct and a diffracted beam. In fact it is very simple to achieve a very good approximation to two beam conditions – at least at 100 kV where the Ewald sphere curves away from the systematic reflections. The two theories, the kinematic theory and the dynamical theory, will be discussed briefly in turn.

5.2 THEORIES OF DIFFRACTION CONTRAST IN TRANSMISSION ELECTRON MICROSCOPY

5.2.1 Kinematic theory of diffraction contrast

Additional assumptions are made in this theory, the most important of which is that the diffracted beam is always weak so that the intensity of the incident beam can be assumed to be unchanged. The method of calculating the intensity at the bottom of each column is to imagine that each column is divided into slabs perpendicular to the diffracted beam and that these slabs can be treated as Fresnel zones. The contribution at the bottom surface from each slab is summed, taking account of phase differences, and for a foil of thickness t this yields an expression for the amplitude of the diffracted beam of

$$\phi_g = \frac{\pi i}{\xi_g} \int_0^t \exp(-2\pi i s z) dz \qquad (5.1)$$

where ξ_g, the extinction distance for the operating reflection **g**, is given by

$$\xi_g = \pi V_c (\cos\theta)/\lambda F(\theta) \qquad (5.2)$$

and V_c is the volume of a unit cell, λ is the wavelength of the electrons, $F(\theta)$ the structure factor for the reflection **g**, and **s** is the deviation parameter.

It **s** is not a function of z then integration of equation (5.1) and multiplication by its complex conjugate to give intensity yields

$$\phi_g^2 = I = \left(\frac{\pi}{\xi_g}\right)^2 \frac{\sin^2(\pi t s)}{(\pi s)^2} \qquad (5.3)$$

Equation (5.3) is plotted in Fig. 5.1 for $s \sim 0$ and for $s \gg 0$ for a crystal of varying thickness and constant **s**. The oscillations in intensity in Fig. 5.1 are

ANALYSIS OF MICROGRAPHS 115

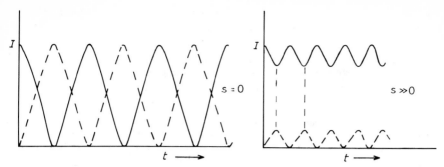

Fig. 5.1 Calculated intensities of the direct (full line) and diffracted beam (broken line) for a crystal of increasing thickness for two values of s using equation (5.3).

referred to as thickness fringes and are analogous to contours on maps since successive fringes occur at equal increments in thickness. The intensity of the direct beam is simply given by $(1 - |\phi_g|^2)$ for unit incident beam intensity.

Neither of these sets of curves agrees quantitatively with observation; thickness fringes are observed but their spacing does not follow the prediction summarized in Fig. 5.1 and they are damped out with increasing thickness. The decreasing intensity of thickness fringes with increase of thickness is caused by inelastic scattering, which is not accounted for in this theory.

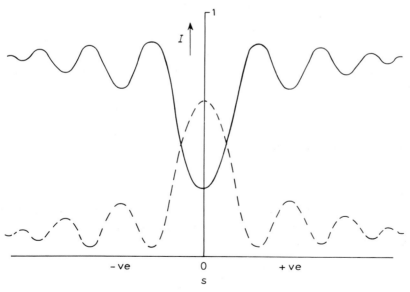

Fig. 5.2 Calculated intensities of the direct (full curve) and diffracted beam (broken curve) for a crystal of constant thickness for values of $w = s\xi_g$ either side of the Bragg condition; calculated from equation (5.3).

Similarly the rocking curves shown in Fig. 5.2 do not agree with observation because inelastic scattering is important in all but very thin crystals. Indeed the kinematic theory is valid only in very thin crystals or when $s_g \gg 0$ since as is evident from Fig. 5.1 the intensity of the diffracted beam exceeds that of the direct beam. At $s = 0$ equation (5.3) reduces to $|\phi_g|^2 = (\pi t/\xi_g)^2$ and thus for crystals with $t > \xi_g/\pi$ the diffracted beam is calculated to be more intense than the incident beam. Note that if s is much greater than zero the coupling between the direct and diffracted beams is reduced and the diffracted beam remains weak. This means that at large deviation parameters the basic premise of the kinematic theory is not violated and the theory can indeed be useful.

The main interest in developing a theory to calculate image information is to calculate the contrast expected from crystal defects and all that needs to be done is to calculate the extra phase change introduced by the displacement due to the strain field of the defect. This phase change associated with the displacement **R** of a unit cell (or one of the slabs in the earlier discussion) enters into the equation for ϕ_g as shown

$$\phi_g = \frac{\pi i}{\xi_g} \int_0^t \exp[-2\pi i (sz + \mathbf{g} \cdot \mathbf{R})] dz \qquad (5.4)$$

The main difficulty in solving this equation, and thus deriving intensities, lies in the expression for **R**, since of course for each slab in each column the value of **R** is likely to be different, and calculable accurately, only if anisotropic elasticity theory is used.

In the case of a screw dislocation of Burgers vector **b** in an elastically isotropic medium the displacements are particularly simple [2] and since the only displacements lie along **b** then it is clear from equation (5.4) that if $\mathbf{g} \cdot \mathbf{b} = 0$ the equation reduces to that of a perfect crystal and the screw dislocation will be invisible, i.e. it will show no contrast. For an edge dislocation with a line direction **u** in an elastically isotropic medium the displacements are more complex and both $\mathbf{g} \cdot \mathbf{b}$ and $\mathbf{g} \cdot \mathbf{b} \wedge \mathbf{u}$ have to be zero for the dislocation to be invisible. Physically, all these statements mean that only if the particular set of diffracting planes remain flat in the presence of the dislocation will the dislocation show no contrast. For other two beam imaging conditions contrast will be observed. It is worth emphasizing that in images obtained using two beam diffraction contrast the image contains information about the planes which are strongly diffracting and about no other planes.

The above invisibility criteria form the basis for the most common method used to determine **b** for dislocations since if two sets of planes (for which the diffracting vectors are $\mathbf{g}_1$ and $\mathbf{g}_2$) are found for which a dislocation is invisible then the direction of **b** is given by $\mathbf{g}_1 \wedge \mathbf{g}_2$. It should be clear that the selection of successive, internally consistently indexed diffracting vectors, is best done using Kikuchi maps (Appendix C). For mixed dislocations*, even in isotropic materials, all planes are distorted to some extent and these dislocations never go completely out of contrast. It is therefore necessary to be able to identify the condition $\mathbf{g} \cdot \mathbf{b} = 0$ even when contrast is observed, both because of this fact and because of the fact that in anisotropic crystals edge and screw dislocations will go out of contrast only if they lie perpendicular to elastic symmetry planes [3]. This aspect and other conclusions derived from image computations will be discussed after reviewing the dynamical theory of image contrast.

5.2.2 Dynamical theory of image contrast

The dynamical theory of image contrast [4] recognizes that the diffracted beam can become very strong, so that rediffraction of it must be considered, and also allows absorption to be incorporated into the two beam equations.

The equations developed, which relate the changes in incident and diffracted intensity to depth in the crystal are (without including absorption) for a perfect crystal

$$\frac{d\phi_0}{dz} = \frac{\pi i \phi_0}{\xi_0} + \frac{\pi i \phi_g}{\xi_g} \exp(2\pi i s z)$$

$$\frac{d\phi_g}{dz} = \frac{\pi i \phi_g}{\xi_0} + \frac{\pi i \phi_0}{\xi_g} \exp(-2\pi i s z)$$

(5.5)

where all the symbols have the same meaning as in earlier equations and ξ_0 (dimensionally a length) is proportional to the atomic scattering amplitude for zero angle and is a measure of the refractive index.

Equation (5.5) can be solved and if $\mathbf{g}$ is not a function of z we obtain

$$|\phi_g|^2 = I = \left(\frac{\pi}{\xi_g}\right)^2 \frac{\sin^2(\pi t \bar{s})}{(\pi \bar{s})^2}$$

(5.6)

where

$$\bar{s} = [s^2 + (1/\xi_g^2)]^{1/2}$$

(5.7)

In this case, the periodicity of the oscillations in intensity is predicted to occur with increase of thickness given by $1/t = [s^2 + (1/\xi_g^2)]^{-1/2}$ so that when $s = 0$ the periodicity is given by ξ_g. Alternatively we can say that the

---

* The contrast from a dislocation is controlled by the values of $\mathbf{g} \cdot \mathbf{b}$, $\mathbf{g} \cdot \mathbf{b}_e$ and $\mathbf{g} \cdot \mathbf{b} \wedge \mathbf{u}$ where $\mathbf{g}$, $\mathbf{b}$ and $\mathbf{u}$ have their usual meanings and $\mathbf{b}_e$ is the edge component of the Burgers vector [3]. If a dislocation is to be invisible all three of these dot products must be zero. This is not possible for mixed dislocations.

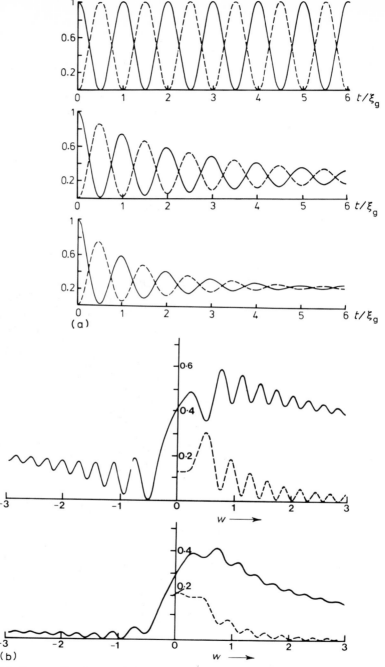

Fig. 5.3 Curves showing the influence of the choice of anomalous absorption parameters on: (a) thickness contours; from top, no absorption, $\xi_g/\xi_g^1 = 0.05$, $\xi_g/\xi_g^1 = 0.1$; and (b) rocking curves; from top $\xi_g/\xi_g^1 = 0.05$, $\xi_g/\xi_g^1 = 0.1$ (from [4]).

effective extinction distance

$$\xi_g^{\text{eff}} = \frac{\xi_g}{(1 + s^2\xi_g^2)^{1/2}} \tag{5.8}$$

In order to match both the observed decay of thickness contours with increasing thickness and the fact that bent crystals do not show complementary contrast in b.f. and d.f., as a crystal is bent through the Bragg condition (see Fig. 5.2) the parameters $1/\xi_0$ and $1/\xi_g$ are replaced by complex quantities. Thus $1/\xi_0$ is replaced by $(1/\xi_0 + i/\xi_0^1)$ and $1/\xi_g$ is replaced by $(1/\xi_g + i/\xi_g^1)$. Values of ξ_g/ξ_g^1 are found empirically by matching experimental and theoretical thickness and bend contours. The influence of the choice of values is illustrated in Fig. 5.3 and it is clear that the fading of thickness contours and asymmetry of bend contours can be generated by equations (5.5) when absorption is included. It is important to note that at symmetry the transmission of electrons is significantly below that at the Bragg condition (Fig. 5.3). This is explicable in terms of the various Bloch waves which are excited [4] and importantly in X-ray microanalysis is associated with enhanced X-ray production for small **g**'s.

The contrast from imperfect crystals appears in the dynamical theory in the same way as in the kinematical theory; a phase factor $2\pi\mathbf{g}\cdot\mathbf{R}$ being introduced into the exponential terms in equations [4].

Rather than go through the derivation of the different forms of the dynamical equations which have been developed for contrast calculations the essential points which emerge from these calculations will be summarized in the following sections. There is in general excellent agreement between theoretical and experimental images and this agreement is presented extremely convincingly in [3].

5.3 ANALYSIS OF IMAGES IN TRANSMISSION ELECTRON MICROSCOPY

5.3.1 Summary of characteristics of images of perfect dislocations

The only possible values for **g·b** for perfect dislocations imaged with fundamental reflections are integers or zero. The parameter **g·b**, although not fully defining the characteristics of images, is a very useful way of describing the imaging conditions. On this basis the following generalizations can be made.

(1) Dislocations are most easily visible and the images broadest when $s = 0$. Images for which $\mathbf{g}\cdot\mathbf{b} \geqslant 2$ show much stronger contrast for the same value of s than those for which $\mathbf{g}\cdot\mathbf{b} = 1$ or 0. At $s = 0$, images for which $\mathbf{g}\cdot\mathbf{b} = 2$ show double images, and as s is increased one of the images

fades away. Images for which $\mathbf{g}\cdot\mathbf{b} = 0$ also show double images at $\mathbf{s} = 0$ but the two peaks fade away together as $\mathbf{s}$ is increased. The images for which $\mathbf{g}\cdot\mathbf{b} = 1$ also fade away as $\mathbf{s}$ is increased and become so much weaker than images for which $\mathbf{g}\cdot\mathbf{b} = 2$ that they can be confused with images for which $\mathbf{g}\cdot\mathbf{b} = 0$. Because the background intensity in dark field becomes very small at large values of $\mathbf{s}$ (cf. Fig. 5.3(b)) weak images can be observed as bright lines on a dark background. At small values of $\mathbf{s}$ the bright and dark field images appear qualitatively similar.
(2) Dislocations running from top to bottom of a foil show oscillatory contrast when imaged at $\mathbf{s} \sim 0$. This contrast is damped out as $\mathbf{s}$ is increased. Images taken in bright field with a given diffracting vector give similar oscillatory contrast to that observed in dark field for the same sense of $\mathbf{g}$ for the part of the dislocation at the top of the foil.
(3) The image of a dislocation lies to one side of a dislocation provided $\mathbf{g}\cdot\mathbf{b} \neq 0$ and $\mathbf{s} \neq 0$. The origin of this can be seen from simple diagrams representing the strain fields around dislocations. The displacements on one side of an edge dislocation can be seen to rotate the crystal towards the Bragg condition (thus giving strong contrast) and on the other side away from this condition. The side to which the image lies is given by the sign of $(\mathbf{g}\cdot\mathbf{b})\mathbf{s}$ and the magnitude of the shift by the magnitude of $(\mathbf{g}\cdot\mathbf{b})\mathbf{s}$. Thus when $\mathbf{g}\cdot\mathbf{b} = 0$ the image is centred on the dislocation. An indication that $\mathbf{g}\cdot\mathbf{b} = 0$, even when this condition gives strong residual contrast can therefore be obtained by reversing $\mathbf{g}$; the absence of an image shift for $|s\xi| > 1$ implies that $\mathbf{g}\cdot\mathbf{b} = 0$.
(4) The image of a dislocation reverses top to bottom and side to side if $\mathbf{g}$ is reversed. A closely spaced dislocation dipole can give apparently more complex behaviour on reversing $\mathbf{g}$ since the images of the dislocations may overlap.

The above summary forms the basis for the analysis of the Burgers vectors of dislocations, the analysis of dislocation loops and small defect clusters and the high resolution dark field weak beam technique. All these techniques will be discussed in the next section.

5.3.2 Burgers vector analysis and loop analysis

The principle underlying the most popular technique for the determination of the Burgers vector of a dislocation has been outlined in Section 5.3.1. It was pointed out that if two diffracting vectors $\mathbf{g}_1$ and $\mathbf{g}_2$ are found for which $\mathbf{g}\cdot\mathbf{b}$ is recognizably zero then $\mathbf{b} = \mathbf{g}_1 \wedge \mathbf{g}_2$. For example, if in a b.c.c. crystal $\mathbf{g}_1 = \bar{1}10$ and $\mathbf{g}_2 = 2\bar{1}\bar{1}$ then $\mathbf{b}$ lies along $[111]$. The main difficulty with this technique lies in the difficulty of recognizing the condition $\mathbf{g}\cdot\mathbf{b} = 0$ for mixed dislocations, for edge dislocations when $\mathbf{g}\cdot\mathbf{b} \wedge \mathbf{u} \neq 0$ and for dislocations in elastically anisotropic crystals. Since weak contrast is observed when $\mathbf{g} \cdot \mathbf{b} = 1$ for large values of $\mathbf{s}$ then $\mathbf{s}$ should be varied in order to determine the origin of weak contrast. Note that if $\mathbf{s}$ is varied and $\mathbf{g}$ is reversed the conditions $\mathbf{g} \cdot \mathbf{b} = 0$, 1 and 2 behave very differently and any possible confusion should be removed; positive recognition of the $\mathbf{g} \cdot \mathbf{b} = 2$ condition is just as valuable as identification of $\mathbf{g} \cdot \mathbf{b} = 0$. As discussed earlier all images are at their strongest when $\mathbf{s} = 0$, and $\mathbf{g} \cdot \mathbf{b} = 0$ and $\mathbf{g} \cdot \mathbf{b} = 2$ images are double. As $\mathbf{s}$ is increased the two peaks of the $\mathbf{g} \cdot \mathbf{b} = 0$ images do not move but both become progressively weaker. For $\mathbf{g} \cdot \mathbf{b} = 1$ images the images become weaker as the value of $\mathbf{s}$ is increased from zero and the image is displaced further from the core. For $\mathbf{g} \cdot \mathbf{b} = 2$ images one of the two peaks weakens as $\mathbf{s}$ is increased and the displacement of the remaining peak from the core increases; if $\mathbf{g}$ is reversed the peak that was weak at large $\mathbf{s}$ becomes strong and a large image shift is therefore observed for large $\mathbf{s}$ when $\mathbf{g}$ is reversed for $\mathbf{g} \cdot \mathbf{b} = 2$ images.

If all the above information is used it is usually possible to define the direction of $\mathbf{b}$ for an isolated perfect dislocation. The magnitude of $\mathbf{b}$ is usually assumed to be the smallest lattice vector in the direction of $\mathbf{b}$. This can be confirmed by selection of $\mathbf{g}$'s which would give $\mathbf{g} \cdot \mathbf{b} = 1$ or 2 for different magnitude $\mathbf{b}$'s and examining the image at $\mathbf{s} = 0$.

A second method of determining the Burgers vector of a dislocation, which is simply a logical extension of the $\mathbf{g} \cdot \mathbf{b} = 0$ technique and of image matching [3], makes use of the detailed nature of the topological/oscillatory contrast visible when $s_g \sim 0$ [5]. A careful assessment of experimental and computed images for crystals that are not highly anisotropic (cf. [3]) has shown that the signs of $\mathbf{g} \cdot \mathbf{b}$ and $\mathbf{g} \cdot \mathbf{b} \wedge \mathbf{u}$ can be obtained from the detailed way in which the intensity varies along a dislocation image. A series of images

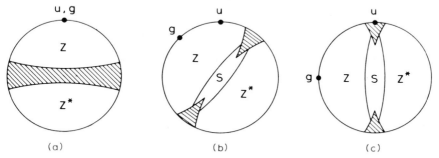

Fig. 5.4 Series of schematic stereograms summarizing the relation between the appearance of images and the value of $\mathbf{g} \cdot \mathbf{b}$ for different possible angles between $\mathbf{g}$ and $\mathbf{u}$. For Burgers vectors within the region defined by Z and Z* zigzag images are observed. For Burgers vectors in the area defined by S and S* (S* is on reverse side of sphere and not marked) discontinuous images with alternate bright and dark maxima are observed and in the hatched regions dislocations will be either invisible or show very weak contrast (see text). (Taken from [5].)

taken with different **g**'s for which the sign of **g**·**b** and **g**·**b** ∧ **u** can be assessed allows immediate identification of the sign and the direction of **b**. The method is best summarized using a series of stereograms which are shown in Fig. 5.4 taken from [5]. These figures define the different types of image observed for various possible angles between **g** and **u**. The beam direction is unimportant and is simply constrained to be perpendicular to **g**. If the angle between **g** and **b** is not too large then the contrast is dominated by the term **g**·**b** (see earlier footnote) and zigzag contrast is observed which corresponds to areas Z or Z^* on Fig. 5.4. If the term **g**·(**b** ∧ **u**), dominates (i.e. **g**·**b** ~ 0) the image becomes discontinuous along its length and appears to be composed of a series of intensity minima and maxima. This is represented by the region S or S^* on the figure. If the dislocation is invisible, or if the image is very weak, then **b** must lie within the shaded regions. The positions of the boundaries between these regions have been defined empirically and cannot be defined precisely. This lack of precision is not important in determining **b** as will be seen below.

Whether the Burgers vector lies within Z or Z^* or S or S^* (note S^* is on the bottom of the stereographic sphere and is therefore not marked) can be determined by the nature of any observed asymmetry in the image. Thus if **g**·**b** > 0 (**g** is acute to **b**) so that **b** lies within region Z rather than Z^* the first dark maximum in a zigzag image on a positive point appears on the left of the dislocation when looking along + **u** (the positive direction is selected arbitrarily) and vice versa if **g**·**b** < 0. If the image is symmetrical side to side then **g**·**b** ~ 0 (see earlier discussion) and **b** lies within S or S^*. If a dark maximum appears on the end of the dislocation at the bottom of the foil **g**·(**b** ∧ **u**) < 0 and **b** lies within area S^* and vice versa for area S. Stereomicroscopy (see Appendix D) is therefore necessary to determine the sign of **g**·(**b** ∧ **u**).

The technique described above allows a rapid assessment of the sign of **g**·**b** (and recognition of the special case **g**·**b** = 0, which is the information used for the invisibility criteria). If two or three **g**'s are used it is usually possible to distinguish between possible candidates for **b** because the first **g** allows the hemisphere which contains **b** to be defined and a second **g** defines a second hemisphere for **b**. Thus **b** must lie within the area common to these two images. If only one possible **b** lies within area then **b** is defined but if not, a third **g** can be used to remove any uncertainty.

An example of this technique is illustrated in Fig. 5.5 and many examples can be seen in [3]. Perhaps it should be emphasized that this technique is empirical in the sense that it has not yet proved possible to predict the symmetry of images. An inspection of the many computed and experimental images available in the literature suggests that it is a well tested technique, although it should be noted that if **B**, the beam direction, and **N** the foil

ANALYSIS OF MICROGRAPHS 123

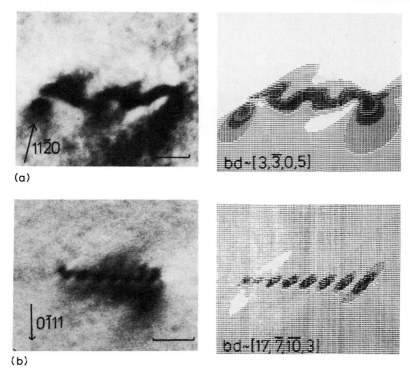

Fig. 5.5 Experimental and computed images of a dislocation in a titanium alloy showing zigzag contrast in (a) with the first dark maximum on the left of the image when viewed from the right side of the micrograph, and a discontinuous image in (b). From these experimental micrographs it therefore follows that $\mathbf{g}\cdot\mathbf{b}$ is positive for (a) and of the possible $\frac{1}{3}\langle 11\bar{2}3\rangle$ dislocations with positive c components only $\frac{1}{3}[\bar{1}2\bar{1}3]$, $\frac{1}{3}[11\bar{2}3]$ and $\frac{1}{3}[2\bar{1}\bar{1}3]$ fulfil this. Image (b) shows that $\mathbf{g}\cdot\mathbf{b} = 0$ and that a bright maximum is at the bottom of the foil shows that $\mathbf{b}$ lies within S rather than S* so that of these possibilities only $\frac{1}{3}[\bar{1}2\bar{1}3]$ fits the observations. (Taken from [25].)

normal, are very different the topological information is not as easy to interpret. The images tend to be sheared so that side to side symmetry of an image is displaced along the length of the dislocation [3]. In order to be confident of the symmetry properties of an image it is sensible to vary **s** and to reverse **g** for all images.

An additional technique to determine **b**, which has the added advantage that the magnitude and sign of **b** are obtained, uses weak beam imaging and this technique is discussed in Section 5.3.3.

The fact that the image of a dislocation generally lies to one side of the core enables the sense of **b**, and hence the vacancy/interstitial nature,

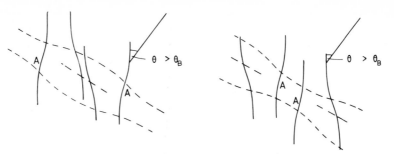

Fig. 5.6 Schematic diagram illustrating the change in image position on reversing **g**. In each case strong contrast will originate from the area marked A since the planes are bent towards the Bragg angle at these points.

to be determined for dislocation loops. Thus the FS/RH convention* shows (e.g. [8]) that a vacancy loop has a component antiparallel to **n** and an interstitial loop has a component parallel to **n**, where **n** is the upward drawn normal to the plane of the loop. If the loops are pure edge then **b** is either antiparallel (vacancy) or parallel to **n** (interstitial).

Interstitial and vacancy edge loops are drawn schematically in Fig. 5.6 and the distortion of the set of planes giving rise to the diffracted beam is shown. Note that it is the distortion of these imaging planes which is relevant when discussing the contrast expected from a loop, and of these planes only; the distortion of the planes which define **n** are irrelevant since the only set of planes influencing the diffracted intensity, when two beam conditions are used, are those planes set near Bragg. Contrast in the image will come from those parts of the diffracting planes which are bent towards the Bragg condition. Since the crystal is imagined to be set with positive **s** then those parts of the diffracting planes marked A will diffract strongly. As shown in Fig. 5.6 these regions lie outside the dislocation core for the interstitial loop and inside for the vacancy loop for the sence of **g** shown; if **g** is reversed the opposite will be true with the strong contrast from the interstitial loop coming now from a region inside the core and coming from outside for the vacancy loop.

The images of loop are thus seen to change size, in opposite senses, as **g** is reversed. This can be summarized by the rule that if (**g·b**) **s** is positive then the image will lie outside the core and if it is negative the image will be inside.

The determination of the nature of an edge loop is then carried out as follows. The Burgers vector is determined by finding g_1 and g_2 for which

---

* If the positive direction around a dislocation loop is taken as clockwise when viewed from above, the Burgers vector of the loop is obtained using the FS/RH convention as follows. A closed clockwise circuit is made looking along the positive sense of the dislocation around any part of the loop beginning and finishing at S and F respectively. An identical circuit is made in a perfect crystal and the closure failure, FS, gives **b**.

$\mathbf{g} \cdot \mathbf{b} = 0$ and hence defining $\mathbf{b}$ by $\mathbf{g}_1 \wedge \mathbf{g}_2 = \pm \mathbf{b}$. Image the loop using a positive $\mathbf{s}$ and reverse $\mathbf{g}$ maintaining $\mathbf{s}$ positive. (It is best to use diffracting vectors for which $\mathbf{g} \cdot \mathbf{b} = \pm 2$ for the image shift experiments.) If we consider loops on $\{100\}$ in irradiated α-iron the following results would be typical in a loop analysis experiment. If images taken with $\mathbf{g} = 1\bar{1}0$ and 200 correspond to $\mathbf{g} \cdot \mathbf{b} = 0$ then $\mathbf{b} = 1\bar{1}0 \wedge 200 = \pm [001]$. If this loop is now imaged with $\mathbf{s}$ positive with $\mathbf{B} = [\bar{1}\bar{1}1]$ and $\mathbf{g} = \pm 112$ and if outside contrast is observed with $\mathbf{g} = 112$ then $(\mathbf{g} \cdot \mathbf{b})\mathbf{s}$ must be positive. $\mathbf{b}$ is therefore $[001]$ and not $[00\bar{1}]$ and since $[001]$ is upwards when $\mathbf{B} = [\bar{1}\bar{1}1]$ the loop is interstitial.

This type of analysis requires that all indexing of electron beam directions, diffracting vectors, etc. are internally self-consistently indexed. This is done using Kikuchi maps (see Appendix C).

If the dislocations are not pure edge then complications arise in the analysis. Several methods are available ([6], [7]) which overcome these problems and two will be discussed in the following section. The origin of the problems which arise when using the technique outlined above are illustrated in Fig. 5.7. Thus Fig. 5.7 shows schematic diagrams in which it is considered that $\mathbf{b}$ has been determined and is therefore fixed in space but $\mathbf{n}$ is allowed to vary since it is unknown. If $\mathbf{n}$ changes from the direction which would correspond to pure edge through the edge-on orientation, i.e. the condition $\mathbf{n} \cdot \mathbf{B} = 0$ (Fig. 5.7) then the FS/RH convention leads to a reversal in the sense of $\mathbf{b}$ and therefore the opposite contrast to that observed for the edge loop will be seen [5]. If $\mathbf{n}$ is rotated still further, through the shear

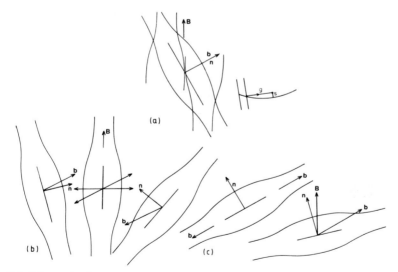

Fig. 5.7 Schematic diagrams illustrating the influence of changing non-edge character on the inside–outside behaviour of a dislocation loop (see text).

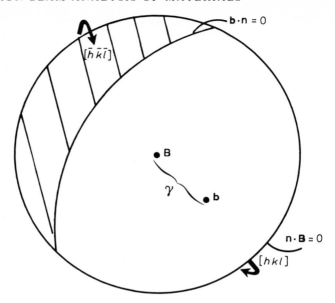

Fig. 5.8 Stereogram centred on **B**, the electron beam direction, showing the orientations of loops for which edge contrast would be observed. The region for which reverse contrast is observed, the unsafe region, is smaller the closer **b** is to **B**, see text (from [6]).

orientation (i.e. beyond **n·b** = 0) then the contrast will revert to that of the edge loop, since **g·b** is now positive again. It therefore follows that edge loop behaviour will be observed for mixed loops when **n** lies within the regions bounded by the conditions **n·B** = 0 and **n·b** = 0. This is summarized in Fig. 5.8.

If **n** cannot be determined accurately (for example if the loop is irregular and/or small) then Fig. 5.8 shows that it is essential that the angle between **B** and **b** be as small as possible. In many crystals it is common practice to assume that the angle between **n** and **b** is defined by some known unfaulting reaction (from an edge to a mixed loop). For example in f.c.c. crystals mixed loops generally form from the unfaulting of $\frac{1}{3}\langle 111 \rangle$ loops to give $\frac{1}{2}\langle 110 \rangle$ loop on $\{111\}$ so that **b** makes an angle of about 35° with **n**. On this basis it is assumed that if the angle between **b** and **B** is less than 55° (i.e. less than $90 - 35°$) then the condition **n·B** = 0 cannot be achieved and edge-type contrast will be observed. Unfaulted loops can of course rotate on their glide cylinder and thus change the angle between **n** and **b** so that it is not always safe to assume that this angle is defined by the unfaulting reaction. In addition unfaulted loops may be observed for which there is no clearly defined unfaulting reaction so that the angle between **n** and **b** cannot be inferred. Thus if non-edge loops are to be analysed with any confidence either the loop normal must be defined or a different technique must be used. It should be noted that loop planes can usually be determined with the appropriate accuracy by noting the shape change on tilting through large angles. If the loop plane can be tilted into the vertical orientation then the **g** normal to the plane accurately defines **n**.

A different method of loop analysis which is in fact best used for loops where the angle between **n** and **B** is large is described below.

The convention used in this method is as follows. The positive direction of **b** of a loop is defined by considering the sense of the movement of the material below the loop to accomodate the point defects in the loop [7]. Thus taking **n** as acute to **B** then $\mathbf{n} \cdot \mathbf{b} > 0$ for a vacancy loop and $\mathbf{n} \cdot \mathbf{b} < 0$ for an interstitial loop. This convention leads to the opposite sense of **b** to that deduced using the FS/RH convention. Thus for $(\mathbf{g} \cdot \mathbf{b})s$ positive an edge loop will now show inside contrast. This difference is obviously a result of choosing **b** of opposite sign but a more important difference arises when considering non-edge loops. Because the sense of the line around the loop does not figure in this convention for defining **b** the contrast for a non-edge loop always conforms to that of an edge loop. The important parameter which needs to be determined experimentally is the sense of the inclination of the loop, i.e. the sense of **n**. This is best done by tilting through a large angle and observing the change in image size for the identical diffraction conditions for a variety of **B**. Having determined the sense of the slope (and it is the sense that is required, not an accurate determination) images are taken using a suitable diffracting vector in $\pm \mathbf{g}$ to determine the sense of image shifts; hence the value of $(\mathbf{g} \cdot \mathbf{b})s$ can be defined and the vacancy/interstitial nature of the loop established.

In principle this technique does not suffer from the complications associated with safe and unsafe orientation for non-edge loops and as such appears to be a more general technique. In practice, however, it is best suited to relatively steeply inclined loops so that the sense of slope is more obvious on tilting, and on this basis the two techniques for loop analysis are complementary; the first method being better suited to loops which are shallowly inclined and the second to steeply inclined loops. This complementary nature can lead to a much more efficient analysis of loops in a given situation. For example considering a b.c.c. metal containing loops of $\mathbf{b} = \frac{1}{2} \langle 111 \rangle$ only those loops of $\mathbf{b} = \frac{1}{2}[111]$ and $\frac{1}{2}[\bar{1}11]$ can be safely analysed with $\mathbf{B} \sim [011]$ if the first technique is used. Assuming the loops were first formed on $\{011\}$ this orientation is safe since **b** is within 35° of **B**. Thus, tilting to $\mathbf{B} = [\bar{1}33]$ and $[133]$ and using the 310 diffracting vectors available in these directions, enables the sense of $(\mathbf{g} \cdot \mathbf{b})s$ to be determined. The other two types of loops of $\mathbf{b} = \frac{1}{2}[1\bar{1}1]$ and $\frac{1}{2}[11\bar{1}]$ can be analysed by tilting away from $[011]$ using $\mathbf{g} = 200$ and hence the sense of the slope of **n** will be immediately defined. Imaging with the appropriate diffracting vectors available with $\mathbf{B} = [021]$ and $[012]$ enables the sign of $(\mathbf{g} \cdot \mathbf{b})s$ to be established.

The two techniques described require that the loop be resolvable so that the inside/outside nature of the images is obvious. This means that the minimum loop diameter which can be analysed is about $\xi_g/3$. Loops smaller than this appear as black spots or as black/white images. There are two methods available for analysis of small loops; either imaging under conditions where ξ_g is decreased or analysing the black/white contrast in detail. Imaging with a reduced value of ξ_g will be discussed first and analysis of black/white contrast next.

5.3.3 Analysis of defects using the weak beam technique

The resolution of diffraction contrast images can be improved significantly simply by increasing the value of **s** since the image width is about $\xi_g^{\text{eff}}/3$ and $\xi_g^{\text{eff}} = \xi_g/(1 + s^2\xi_g^2)^{1/2}$ (Section 5.2.2). As discussed earlier the intensity of images taken at large values of **s** is low and in bright field these weak images are not easily seen since the background intensity remains high (cf. Fig. 5.2). In dark field however, the background intensity is very low for large values of **s** and weak images therefore give significant contrast. Thus images taken at large **s** in dark field will give weak, but high contrast, narrow images of dislocations, with an effective improvement in resolution of a factor 5 to 10. For example, if the two beam extinction distance is taken as 40 nm, and s_g as 0.2 nm^{-1} then from the above equation $\xi_g^{\text{eff}} = 5.0$ nm. The extinction distance has been reduced by a factor of eight, and this leads to a corresponding reduction in the width of dislocation images. This technique enables the loop analysis methods discussed in the previous section to be applied to loops down to about 5 nm in diameter.

Very high contrast is observed in weak-beam images and the origin of this high contrast can be seen from the way in which diffraction contrast arises. The term in the dynamical equations which includes **R** can be written as $(\mathbf{s} + \mathbf{g} \cdot d\mathbf{R}/dz)$ (cf. [4]) and this form of the equation makes it clear that the origin of contrast lies in the local changes in **s** caused by the term $\mathbf{g} \cdot d\mathbf{R}/dz$. When the term $(\mathbf{s} + \mathbf{g} \cdot d\mathbf{R}/dz) = 0$ the crystal is locally at the Bragg condition and there is therefore efficient coupling between the direct and diffracted beams (cf. Fig. 5.1). If, as in weak beam, **s** is set very large then the transfer of energy from the direct to the diffracted beam will be limited (cf. Fig. 5.1) and only in regions very close to the core of the dislocation will $\mathbf{g} \cdot d\mathbf{R}/dz$ be large enough to satisfy the condition $(\mathbf{g} \cdot d\mathbf{R}/dz + \mathbf{s}) = 0$ and efficient energy transfer take place.

Because weak beam images are not easily seen on the screen it is useful to make use of the dark field facilities (which allow rapid switching to and from dark field) to carry out the focussing in bright field. This can be done very easily by setting the sample at Bragg for the appropriate **g** in bright field, focussing and then tilting **g** (not **ḡ**) down the optic axis using the d.f.

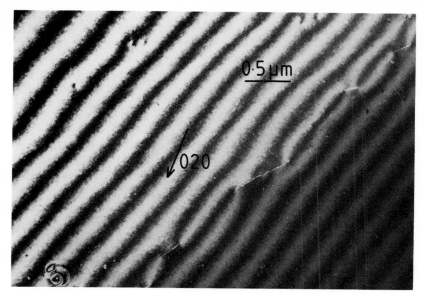

Fig. 5.9 Weak beam micrograph taken with $\mathbf{g} = 020$ from a f.c.c. alloy showing one terminating thickness contour at the dislocation (see text).

tilt coils. When $\mathbf{g}$ is down the optic axis $3\mathbf{g}$ will be at Bragg* and this corresponds to a very large value of $\mathbf{s}$ for $\mathbf{g}$. If the images in this weak beam dark field image are not sufficiently narrow return to b.f., tilt the crystal to a positive $\mathbf{s}$, refocus and return to d.f. Because the d.f. coils have not been touched $\mathbf{g}$ will still be axial but $\mathbf{s}$ will now be even larger. Successive adjustments enable optimum weak beam conditions to be achieved. It has been found that $\mathbf{s}$ must be greater than about $2 \times 10^{-2} \, \text{Å}^{-1}$ for satisfying weak beam imaging.

As mentioned in Section 5.3.2 the weak beam technique can be used to determine the direction and magnitude of $\mathbf{b}$. The contrast effect used in this technique is illustrated in Fig. 5.9 where it can be seen that one of the closely spaced thickness fringes terminates at either end of the dislocation. It can be shown (e.g. [9]) that the number of fringes n, which terminate on a dislocation is given by $\mathbf{g} \cdot \mathbf{b}$, and if successive weak beam images are taken with three non-coplanar diffracting vectors $\mathbf{b}$ can therefore be determined by noting the value of n for each value of $\mathbf{g}$; the sign of n can be taken arbitrarily as say positive if the terminating fringe is on one specific side of the dislocation. For example if a dislocation in molybdenum were imaged with $\mathbf{g} = \bar{1}2\bar{1}, 1\bar{1}0$ and $01\bar{1}$ the following deductions could be made:

* It is left to the reader to show this to be the case (see [8]).

g	n	**g·b**	
$\bar{1}21$	2	$\frac{1}{x}[-h-2k+1] = 2$	
$01\bar{1}$	1	$\frac{1}{x}[k-1] = -1$	therefore $\mathbf{b} = \frac{1}{2}[\bar{1}\bar{1}1]$
$1\bar{1}0$	0	$\frac{1}{x}[h-k] = 0$	

This technique requires that thickness contours should be sufficiently closely-spaced to allow assessment of the number of terminating fringes; low order reflections and weak beam conditions are therefore required. It should be noted that despite the fact that low order diffracting vectors are used high index Burgers (which may be present in grain boundaries) can be identified. This fact makes this technique very valuable.

5.3.4 Contrast from voids and precipitates

The contrast from large voids is influenced by the diffracting conditions. Firstly the projected shape is determined by **B**, the electron beam direction. Secondly, if the void is faceted, then in some directions the void will produce thickness fringes since the inclined face of the void will effectively produce a tapered crystal. Again therefore the presence of the fringes will be a function of **B** but in addition will also be controlled by the values of **g** and **s** since these determine ξ_g, the extinction distance, and thus the fringe spacing.

For smaller voids the possibility of observing fringes is precluded when the projected void thickness is less than the effective extinction distance. If small voids are imaged in a thin crystal ($\leqslant 3\xi_g$) near the condition $\mathbf{s}_g = 0$ when strong thickness fringes will be observed, the voids which are regions where the overall thickness of the crystal will be decreased locally, will show contrast. Voids will appear dark at the side of a bright fringe where a decrease in thickness would lead to a dark fringe but will appear bright on the edge of a dark fringe where a decrease in thickness would lead to an increase in the intensity of the beam used to form the image. In thick crystals ($\geqslant 6\xi_g$) voids will always be brighter than background when $\mathbf{s} \sim 0$.

When small voids are imaged when $\mathbf{s}_g \neq 0$ they can show contrast above or below background since a phase shift of magnitude $2\pi \mathbf{s}_g t$ (where t is the void size in the beam direction) occurs in the diffracted beam [4]. The contrast observed is then a function of the position of the void in the foil. The contrast from very small voids, imaged in bright field using a very large value of $\mathbf{s}_g$, is found to be a very sensitive function of the degree of under or overfocus of the objective lens. The voids show minimum contrast at focus. In under-focussed images the images are above background but are surrounded by a dark ring whereas in overfocussed images the images show contrast below background intensity. Calculations [10] show that the true size of the voids is obtained by measuring the inner diameter of the annular dark ring in the underfocussed image. Through-focus images are very useful both in identifying voids and obtaining their true size.

Very small voids may have significant elastic strain associated with them in which case the contrast is similar to that expected from a spherical inclusion which has a misfit with the matrix.

The contrast observed from a misfitting precipitate is a function of the precipitate size and the matrix strain as well as of the diffracting conditions and the position of the precipitate in the foil. Calculations have shown [11] that the contrast is controlled by the parameter P, given by

$$P = g\Delta V / \xi_g^2 \pi \tag{5.9}$$

where ΔV is the volume misfit of the defect and is given by $\Delta V = \pi b R^2$ for a dislocation loop of radius R and $\Delta V = 4R^3 \Sigma / 3$ for a spherical inclusion of radius R and volume misfit Σ.

If a precipitate reflection lies within the objective aperture, so that interference can occur between the direct beam and the precipitate reflection, or, if dark field is being used, between a matrix diffracted beam and a precipitate reflection, Moiré fringes will be generated (e.g. [4]). If the spacing of parallel matrix and precipitate planes which are giving rise to the two reflections are d_1 and d_2 then the spacing D of the Moiré fringes is given by

$$D = \frac{d_1 d_2}{|d_1 - d_2|} \tag{5.10}$$

If the planes have identical spacings but the precipitate is rotated α with respect to the matrix then a Moiré of spacing D_r given by

$$D_r = \frac{d}{2 \sin (\alpha/2)} \tag{5.11}$$

is formed.

If a rotation and a difference in spacing are both present then the Moiré spacing D_m is given by

$$D_m = d_1 d_2 / [(d_1 - d_2)^2 + d_1 d_2 \alpha^2]^{1/2} \tag{5.12}$$

5.3.5 Analysis of black/white contrast from small and large centres of strain

The contrast from small centres of strain ($P \ll 1$), including a vacancy or interstitial loop, at $s_g \sim 0$ consists of black/white lobes and for precipitates

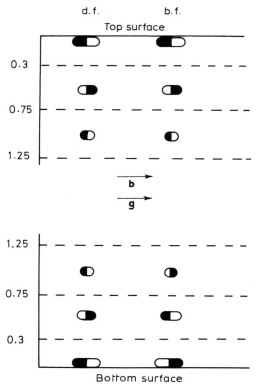

Fig. 5.10 Schematic diagram illustrating the depth dependence of black/white contrast observed for small strain centres with edge loop strain field in an elastically isotropic crystal.

or loops with edge dislocation displacements in an elastically isotropic material the black/white vector **l** is approximately parallel or antiparallel with **g**. Whether **l** lies parallel or antiparallel to **g** is a function of the depth of the object in the foil and whether the defect has vacancy or interstitial strain field. Computed images for b.f. and d.f. are summarized in Fig. 5.10. (Images taken when $s_g \neq 0$ simply appear as black dots, the contrast becoming progressively weaker as s_g is increased.) Clearly therefore the distinction between very small clusters with interstitial or vacancy strain fields requires that the distance of the defect from the foil surface be accurately determined and that images be obtained at $s_g \simeq 0$. In order to measure the distance that defects are from the foil surface it is best to decorate one (defined) surface with gold islands and then to use stereomicroscopy (see Appendix D) to measure the vertical distance between the gold and the defects. The presence of a polishing layer or an oxide layer between the gold and the foil surface

ANALYSIS OF MICROGRAPHS 133

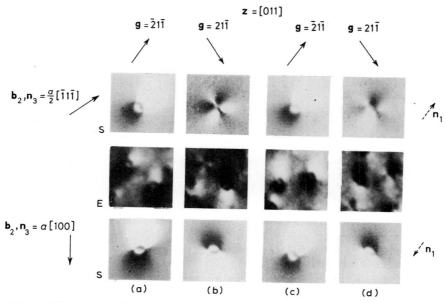

Fig. 5.11 Experimental and computed images for mixed and edge defects in Mo. (after [12]).

can lead to errors when using this technique; the presence of a region with no visible defects may be indicative of the presence of an oxide layer.

If the Burgers vectors of the small loops giving rise to black/white contrast are required, rather than simply a distinction between a vacancy or an interstitial strain field, then it is necessary to examine the observed strain contrast in considerable detail. This is a field where image computation, presented not as profiles but as two-dimensional images [3], has been found to be essential [12].

As in the section dealing with dislocation images it is useful to consider images in terms of the possible values of **g·b**. Thus for an elastically isotropic crystal it has been shown by calculation and by experiment that the following generalizations can be made concerning the black/white images observed when $s \sim 0$. The various cases which are discussed below are illustrated in Fig. 5.11.

(a) Images for which g.b = 0

These images fall into two distinct categories. Firstly, if the angle between **B** and the loop normal is less than 45°, images for which **g·b** = 0 tend to give either zero contrast or weak contrast showing weak black/white lobes. Secondly if the angle between **B** and **n** is 90° then characteristic butterfly images are observed (Fig. 5.11).

(b) Images for which g.b ⩽ 1 but not equal to zero

These images give simple black/white lobes but there is no simple relationship between the direction of **b** and the vector from black to white (BW) in the lobe. Indeed the black/white vector is more sensitive to **g** than to **b**. Contrary to earlier conclusions based on image profiles rather than computed images the direction BW cannot be used to identify the precise direction of **b**.

(c) Images for which g.b > 1

Complex black/white lobes are observed where the interface between the black and white lobes is a region of complex contrast.

These various classifications of possible images allow **b** to be determined in a systematic way as illustrated in Fig. 5.11.

The layer structure (cf. Fig. 5.10) is separated by complex images and any defects in the region between layers are best analysed by using **g**'s with sufficiently different extinction distances so that some of the images are recognizably within a layer.

(d) Images of large centres of strain

The contrast from large centres of strain is more complex than that associated with small strain centres. The oscillatory black/white contrast observed for small strain centres is suppressed and unless the strain centre is close to a surface the contrast under two beam conditions takes the form of large black/white lobes and in dark field the sense of the vector black to white is the same for strain centres near the top and bottom foil surfaces. The sense of the vector black/white then depends only on the interstitial vacancy nature of the strain field and stereomicroscopy is therefore not required in this case. If the strain centre is interstitial the contrast on a positive dark field print is such that the vector BW is parallel to **g** whereas it will be antiparallel for a vacancy defect. This conclusion is based on calculations assuming elastic isotropy [11].

If the only information which is required is the number and size of precipitates or loops in a sample then two beam imaging at $s \simeq 0$ may not be the appropriate imaging mode. Thus if the density of precipitates is very high then the complex images at $s = 0$ will overlap, making it impossible to count individual precipitates. Imaging using a precipitate reflection will generally not reveal all the precipitates since several equivalent orientations may be present. An alternative technique involves using many beam imaging along a prominent zone axis of the matrix. Under these conditions it has been found experimentally, and confirmed by image computing, that clear images, with no long range strain contrast, are obtained [13].

5.3.6 Images of planar defects

Planar defects give rise to contrast in TEM because the change either in interplanar spacing or in stacking sequence associated with the planar defect can introduce discontinuities into planes which intersect the defect. This discontinuity will give rise to contrast, if such planes are the diffracting planes, since a phase change $2\pi \mathbf{g} \cdot \mathbf{R}$ will be introduced by this discontinuity, where $\mathbf{R}$ is the displacement associated with the planar defect and $\mathbf{g}$ is the diffracting vector; electrons scattered by the crystal below the planar defect will suffer a phase change of $2\pi \mathbf{g} \cdot \mathbf{R}$ with respect to electrons scattered by the crystal above the fault. The convention used in earlier work [4] that the bottom part of the crystal is displaced by $\mathbf{R}$ relative to the stationary top of the crystal, will be followed here.

Contrast calculations (e.g. [4]) have shown that only four distinct values of $2\pi \mathbf{g} \cdot \mathbf{R}$ need be considered, corresponding to the conditions $\mathbf{g} \cdot \mathbf{R} = 0$, $\mathbf{g} \cdot \mathbf{R} = $ fraction and $\mathbf{g} \cdot \mathbf{R} = n$, where n is an integer and the special case when $\mathbf{g} \cdot \mathbf{R} = 1/2$. Thus referring to Fig. 5.12(a), which corresponds to the condition $2\pi \mathbf{g} \cdot \mathbf{R} = 0$, it can be seen that there will be no contrast since the displacement $\mathbf{R}$ is perpendicular to $\mathbf{g}$; this displacement does not give rise to a discontinuity in the diffracting planes. The fact that no contrast is associated with this imaging condition is evident from equation (5.4) since this clearly reduces to the equation for a perfect crystal when $2\pi \mathbf{g} \cdot \mathbf{R} = 0$.

Similarly Fig. 5.12(b) shows that if $2\pi \mathbf{g} \cdot \mathbf{R} = 2\pi n$ then again no contrast will be observed since this condition simply means that the component of $\mathbf{R}$ along $\mathbf{g}$ is an integral number of interplanar spacings.

When $2\pi \mathbf{g} \cdot \mathbf{R}$ is fractional (see Fig. 5.12(c)), contrast calculations show that the contrast observed from inclined planar defects takes the form of alternate dark and light fringes which are parallel to the trace of the defect in the foil surface. Fig. 5.12(c) shows that when $\mathbf{g} \cdot \mathbf{R}$ is fractional the diffracting planes are discontinuous and it is then not surprising that contrast is observed. The details of the contrast which is observed when imaging planar defects under strong two beam conditions can be summarized as follows.

If $-\pi < 2\pi \mathbf{g} \cdot \mathbf{R} < +\pi$ so that $\mathbf{g} \cdot \mathbf{R}$ lies between $+1/2$ and $-1/2$ (but does not equal $+1/2$ or $-1/2$) then the fringes observed will be symmetrical about the centre of the foil in bright field but asymmetric in dark field; the fringe corresponding to the intersection of the plane of the defect with the

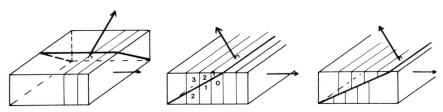

Fig. 5.12 Schematic diagrams showing the origin of contrast observed for stacking faults; (a) $\mathbf{g} \cdot \mathbf{R} = 0$, which will show no contrast; (b) $\mathbf{g} \cdot \mathbf{R} = $ integer, which will also show no contrast; and (c) $\mathbf{g} \cdot \mathbf{R} = $ a fraction, which will therefore show contrast. Horizontal arrow is $\mathbf{g}$.

top surface of the foil appears identical in both bright and dark field, but the fringe at the bottom surface reverses in contrast for the same sense of **g**. The number of fringes is determined by the effective extinction distance and the foil thickness. In very thin crystals extra fringes appear at the foil surfaces as the thickness is increased but as the thickness increases further to $\lesssim 6\xi_g$ extra fringes are now generated in the centre of the foil and the outermost fringes remain either dark or bright depending on the value of **g** and **R**. This property of the outermost fringe enables faults of different values of **R** to be distinguished provided **R** does not differ by a lattice vector, since **g·R** for a lattice vector is always integral; as discussed earlier when $2\pi\mathbf{g}\cdot\mathbf{R} = 2\pi n$ the contrast equations are unchanged so that the addition of a lattice vector cannot alter contrast from a planar defect. This is important when considering contrast from stacking faults in h.c.p. crystals as discussed later.

It follows from this discussion that when comparing experimental and computed images, only those values of $2\pi\mathbf{g}\cdot\mathbf{R}$ which are made as close to zero as possible (by adding or subtracting $2\pi n$) need be considered when assessing the intensity of the outermost fringe to determine the sense of **R**. the outermost fringe will be of the same intensity for values of $+4\pi/3$ and $+\pi/3$ (since these values differ by π) but will be of different intensity from images which correspond to the condition $2\pi\mathbf{g}\cdot\mathbf{R} = -\pi/3$ or $+2\pi/3$. Note that these last two values differ by π and will therefore give identical contrast despite the opposite sign of $2\pi\mathbf{g}\cdot\mathbf{R}$.

The intensity of the fringe contrast observed is determined by the magnitude of $2\pi\mathbf{g}\cdot\mathbf{R}$ (but note again that the value of $2\pi\mathbf{g}\cdot\mathbf{R}$ must be made as close to zero as possible by subtracting (or adding) $2\pi n$ to the value of $2\pi\mathbf{g}\cdot\mathbf{R}$ before assessing the significance of the observed contrast). When $2\pi\mathbf{g}\cdot\mathbf{R} \lesssim 0.01$ the contrast becomes very weak and if weak fringes are to be studied then it is essential to image close to $s_g = 0$ and to defocus the condenser system in order to obtain as parallel illumination as possible. A convergent beam will tend to reduce contrast since the specimen will be imaged simultaneously over a range of values of s_g giving rise to a range of values of ξ_g^{eff}. These points are important when analysing the nature of stacking faults as discussed below. A special case arises when $\mathbf{g}\cdot\mathbf{R} = n/2$, i.e. if $2\pi\mathbf{g}\cdot\mathbf{R} = \pm\pi$. The fringe characteristics differ very significantly from the more general case discussed above and can be summarized as follows:

(a) **At s = 0**

(1) The bright and dark field images for the same **g** are complementary and since the central fringe is bright in bright field it is dark in dark field.
(2) The fringe spacing is $\xi_g/2$ and the intensity of the outermost fringe is determined by the foil thickness; in bright field the fringes formed

at the intersection of the planar defect and the foil surface are dark if the thickness is $n\xi_g$ and bright if the thickness is $(n+\frac{1}{2})\xi_g$, where n is an integer. The new fringes associated with an increase in thickness are formed at the surface.

(b) At $s \neq 0$

The fringe behaviour is now complex and the following illustrates the changes from the $s = 0$ images:

(1) The bright and dark field images are no longer complementary, the dark field image no longer being symmetrical about the centre of the foil.
(2) The spacing of the fringes increases from $\xi/2$ and extra fringes may now be formed at the centre of the foil.

The characteristics of planar defects summarized above form the basis of the analysis of stacking faults in crystals and the following sections deal with analysis of faults (including antiphase boundaries) in f.c.c. and h.c.p. crystals

5.3.7 Fault analysis in f.c.c. crystals

The hard sphere description of the displacement associated with stacking faults, which are formed either by the insertion of a plane of interstitials or by the formation of a plane of vacancies on $\{111\}$, is of the type $\frac{1}{3}\langle 111 \rangle$. Similarly, stacking faults formed by glide of partial dislocations on $\{111\}$ are characterized by displacements of the type $\frac{1}{6}\langle 112 \rangle$. If the fault is on (111) then the fault formed by point defects has $\mathbf{R} = \frac{1}{3}[111]$. The $\frac{1}{3}\langle 111 \rangle$ faults differ from the $\frac{1}{6}\langle 112 \rangle$ faults by $\frac{1}{2}\langle 110 \rangle$, i.e. by a lattice vector, e.g.

$$\tfrac{1}{3}[111] = \tfrac{1}{6}[\bar{1}\bar{1}2] + \tfrac{1}{2}[110]$$

Since $\mathbf{g} \cdot \mathbf{R}$ must equal an integer or zero for lattice vectors it follows (see previous section) that the same contrast will be observed for $\frac{1}{3}\langle 111 \rangle$ faults as for the corresponding $\frac{1}{6}\langle 11\bar{2} \rangle$ faults. It therefore does not matter which displacement is used when discussing contrast from a stacking fault.

Two different types of fault exist in f.c.c. crystals: extrinsic and intrinsic. Formally the distinction between an extrinsic and an intrinsic fault is in the sense of the displacement $\mathbf{R}$. Thus if a stacking fault is on an inclined plane the sense of movement of the bottom part of the crystal to accomodate a plane of vacancies (an intrinsic fault) would be upwards (i.e. acute to $\mathbf{B}$) and downwards to accommodate a plane of self-interstitials (an extrinsic fault). Using this description of stacking faults the calculated and observed contrast can be used to distinguish between these two types of fault. Thus calculations [3] using the dynamical equations have shown that, in a sufficiently thick crystal ($\gtrsim 6\xi_g$), the intensity of the outermost fringe is determined by the value of **g·R** modulo $1/3$; if **g·R** $= +1/3$ then the outermost fringes on a bright field image will be brighter than background and if **g·R** $= -1/3$ they will be darker. Thus if **g** is known and the upward normal to the $\{111\}$ plane is also known then the value of **R** can be determined since **g·R** is known from the intensity of the outermost fringe.

In order to ensure that the outermost fringe is clearly visible, images should be obtained at $s_g = 0$, **g** should be reversed (and hence the outermost fringe should be observed to reverse in intensity) and dark field images should be taken (and hence, for the same sense of **g** in b.f. and d.f. the fringe at the bottom of the foil should be seen to reverse in contrast). Since no contrast is expected when **g·R** $= 0$ or integer, fault analysis follows a similar pattern to the analysis of other defects as the example considered below will show.

Fig. 5.13 shows images of stacking faults in silicon and the planes on which the faults lie have been labelled. The faults labelled E are extrinsic and those labelled I, intrinsic. Note the strong contrast shown by the faults on $(\bar{1}11)$ and $(\bar{1}\bar{1}1)$ for which **g·R** $= +1/3$ and $-1/3$. For the other two faults **g·R** $= \pm 1$ and on the basis of Fig. 5.12 should show no contrast.

The fact that some contrast is observed for $\mathbf{g} = \bar{3}11$ when **g·R** is 1 (on a hard sphere model) but none is observed when **g·R** $= 0$, is indicative that the

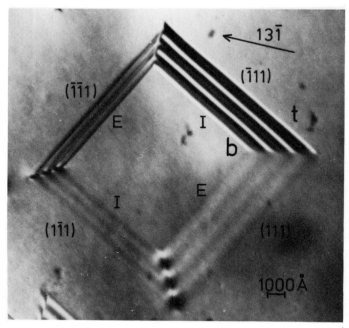

Fig. 5.13 Bright field transmission electron micrographs of stacking faults in silicon imaged near $\mathbf{B} = [2, 3, 11]$ using $\mathbf{g} = 13\bar{1}$. The contrast observed from the faults on the four $\{111\}$ is discussed in the text.

displacement across the fault is not precisely $\frac{1}{3}\langle 1\bar{1}1 \rangle$, i.e. the hard sphere approximation is inadequate ([14], [15]). It is of course not surprising that there is a small supplementary displacement in view of the incorrect stacking sequence associated with a fault. As mentioned earlier it is essential to use imaging conditions which are suitable to detect such a small displacement – $\mathbf{s}_g = 0$ and defocussed illumination. In addition, the optimum thickness for such observations corresponds to thicknesses of $(n + \frac{1}{2})\xi_g$ in bright field and of $n\xi_g$ in dark field where n is an integer.

The specific plane on which a defect lies can be obtained either as discussed above by putting the fault out of contrast, or by defining directions contained in the fault plane. The technique of trace analysis which allows true directions to be obtained from the projected directions in micrographs is explained in Appendix D. The sense of slope of a fault is indicated by comparing dark and bright field images.

5.3.8 Fault analysis in h.c.p. crystals

Although there is evidence that faults are stable on planes other than (0001) in some hexagonal metals and direct evidence for faults on $\{10\bar{1}0\}$ in tungsten carbide for example, the discussion in this section will be restricted to faults on (0001). No new principle emerges when considering faults on other planes, although if the other planes are not stacked ABAB the faults can be very different from those on (0001).

As with f.c.c. metals and alloys both extrinsic and intrinsic faults can be formed either by the condensation of point defects or by glide. Because (0001) planes are stacked ABAB ... in hexagonal metals there are some important differences between faults in hexagonal and f.c.c. crystals and the structure of possible faults will be considered briefly before discussing their analysis.

The shear to produce faults on (0001) is of the type $\frac{1}{3}\langle 10\bar{1}0 \rangle$ but unlike f.c.c. faults (where shear in the opposite senses leads to extrinsic* and intrinsic faulting) opposite sense shears occur on alternate planes in order to produce simple intrinsic faulting. The consequence of this is that, in contrast to the case with f.c.c. metals, the intensity of the outermost fringe when imaged with a diffracting vector $\mathbf{g}$ varies for crystallographically equivalent glide faults since $\mathbf{g} \cdot \mathbf{R}$ changes sign. For example for $\mathbf{R} = \frac{1}{3}[10\bar{1}0]$ or $\frac{1}{3}[\bar{1}010]$ (which on adjacent planes will produce equivalent faults) if $\mathbf{g} = 01\bar{1}0$ is used then $\mathbf{g} \cdot \mathbf{R} = +1/3$ and $-1/3$ respectively. Similarly intrinsic (i.e. vacancy) and extrinsic (i.e. interstitial) faults cannot be distinguished because they differ by a lattice vector [0001] or $\frac{1}{3}\langle 11\bar{2}3 \rangle$. Thus an intersti-

* The shear to produce extrinsic faulting in f.c.c. crystals actually involves the movement of a double partial but the sum of these two $\frac{1}{6}\langle 112 \rangle$ partials is equal and opposite to the third $\frac{1}{6}\langle 112 \rangle$ partial on that $\{111\}$.

tial loop of $\mathbf{b} = \frac{1}{2}[0001]$ differs from the unsheared $\frac{1}{2}[000\bar{1}]$ vacancy loop by $[0001]$, and the sheared interstitial loop of $\mathbf{b} = \frac{1}{6}\langle 20\bar{2}3 \rangle$ differs from a sheared vacancy loop of $\mathbf{b} = \frac{1}{6}\langle 20\bar{2}\bar{3} \rangle$ by either $\frac{1}{2}[0001]$ or $\frac{1}{3}\langle 11\bar{2}3 \rangle$ depending upon the specific $\frac{1}{3}\langle 10\bar{1}0 \rangle$ shear associated with the loops. The term $2\pi\mathbf{g}\cdot\mathbf{R}$ differs for vacancy and interstitial loop only by $2\pi n$, where n is an integer, for all values of $\mathbf{g}$ and hence the fault contrast is identical. Clearly though the sense of $\mathbf{b}$ of the surrounding dislocation is different for a vacancy and an interstitial loop.

The analysis of the faults described above is carried out in a similar manner to that for f.c.c. crystals. Imaging with diffracting vectors for which $\mathbf{g}\cdot\mathbf{R}$ is zero or an integer will lead to the value (plus or minus a lattice vector) for $\mathbf{R}$. For example imaging a fault which has a displacement of $\frac{1}{6}[20\bar{2}3]$ with $\mathbf{g} = 1\bar{2}10$ and $11\bar{2}2$ will yield images for which $\mathbf{g}\cdot\mathbf{R} = 0$. Note that imaging with $\mathbf{g} = 11\bar{2}\bar{2}$ in this example would give an image for which $\mathbf{g}\cdot\mathbf{R} = 2$ and thus reveal any supplementary displacement associated with the fault.

There is a further important difference between fault analysis in f.c.c. and hexagonal crystals which arises when considering the differences between faults formed by climb (i.e. by point defects) and faults formed by shear. In the case of f.c.c. crystals these differ by $\frac{1}{2}\langle 110 \rangle$ (as discussed in Section 5.2.6), a perfect lattice vector, and are therefore indistinguishable. In the case of hexagonal metals the $\frac{1}{2}[0001]$ and $\frac{1}{6}\langle 20\bar{2}3 \rangle$ climb faults differ from the $\frac{1}{3}\langle 10\bar{1}0 \rangle$ glide faults by vectors containing $\frac{1}{2}[0001]$ which is not a lattice vector, so that the faults are different and distinguishable. The distinction can be made by imaging the fault with any $10\bar{1}0$ reflection and with the $10\bar{1}1$ reflection which differs only in the last integer. Images obtained using such diffracting vectors will yield images with the same intensity outermost fringe for a $\frac{1}{3}\langle 10\bar{1}0 \rangle$ fault, since $\mathbf{g}\cdot\mathbf{R}$ will not change sign, whereas if the $\frac{1}{6}\langle 20\bar{2}3 \rangle$ fault gives $\mathbf{g}\cdot\mathbf{R}$ of $+1/3$ for the $10\bar{1}0$ reflection it will give $-1/6$ the $10\bar{1}1$ reflection and the outermost fringe will reverse in intensity.

There is one additional type of stacking fault which may occur on (0001) in h.c.p. crystals which is perhaps best termed a stacking error. This defect consists of just one plane stacked incorrectly so that the stacking sequence is ABABCBAB.... To a first approximation this should give no contrast since there is no phase change but there may be contrast visible if the incorrect stacking leads to a change in interatomic spacing. Thus, as with stacking faults imaged with $11\bar{2}2$ diffracting vectors for which the hard sphere approximation suggests there will be no contrast, contrast may be observed due to the supplementary displacement.

5.3.9 Analysis of antiphase boundaries

Antiphase domains in ordered alloys are displaced with respect to each other by a vector $\mathbf{R}$ whereby atoms of type A are on α-sites on one side of a boundary and on β-sites on the other side. **R** is then the vector joining α and β within a single domain. In a CsCl structure for example this vector is $\frac{1}{2}\langle 111 \rangle$.

The visibility of antiphase boundaries (APBs) is governed of course by the value of **g**·**R** as discussed earlier. Clearly since **R** for APBs is a lattice vector in the disordered crystal the value of **g**·**R** for all fundamental reflections will be integers and no contrast would be expected on a hard sphere model. If superlattice reflections are used to image a specimen containing APBs **g**·**R** may be a fraction and fringe contrast will therefore be observed.

The determination of **R** for APBs follows exactly the same course as the analysis of stacking faults in crystals with the important difference that superlattice reflections must be used and the images are therefore best in dark field. In a Cu_3Au structure, images taken with an 001 and a 110 reflection would put an APB with $\mathbf{R} = \frac{1}{2}[\bar{1}10]$ out of contrast and it is therefore straightforward in principle to determine **R**. The plane on which the APB lies can be determined using trace analysis (see Appendix D). Analysis of APBs is not always straightforward since there may be no suitable superlattice reflections for which **g**·**R** = 0 – as is the case for CsCl alloys.

Another problem in imaging APBs is that the extinction distance of superlattice reflections is determined by the difference in the scattering factor of the two atoms making up the alloy. The extinction distances are therefore larger than those of fundamental reflections and the contrast from APBs commonly consists of a single dark or bright fringe. Weak beam imaging can of course increase the number of fringes and indeed weak beam imaging has been used to image APB tubes which show virtually zero contrast under normal dark field imaging conditions [16].

Because A—A bonds exist across APBs there is likely to be a supplementary displacement in addition to the displacement $\alpha \to \beta$ of the hard sphere approximation. This contrast would be expected to be most significant in alloys where the misfit between the atoms is largest. The contrast expected from an APB which has a significant additional displacement will be most easily observed using a fundamental relfection so that the contrast arises only from the supplementary displacement.

5.3.10 Images of partial dislocations

By definition, partial dislocations border stacking faults in crystals and the contrast observed is influenced significantly by the contrast associated with the fault. This makes it difficult to summarize the contrast from partial dislocations and analysis of partials is best carried out when the fault is out of contrast. Indeed the only way to define the total displacement associated with a fault is to determine the Burgers vector of the associated partial dislocation, the uncertainty in the fault displacement arising because the addition or subtraction of a lattice vector does not change the observed fault contrast.

The smaller the value of **g·b** the smaller the contrast observed in general, but equal values of **g·b** can give images of very different intensity if the fault is in contrast. Thus if the image of the partial is displaced so as to lie outside the image of the fault, weak dislocation contrast will be observed if **g·b** is small but if the image is displaced to lie on the fault image, very strong dislocation contrast may be observed even for small values of **g·b**. This forms the basis for a method for analysis of the nature of faulted loops. Thus, images of a $\frac{1}{3}\langle 111 \rangle$ dislocations obtained with a 200 type diffracting vector and a positive value of **s** show very intense contrast when **g·b** $= -2/3$ (using the FS/RH convention to define **b**) and very weak contrast when **g·b** $= +2/3$. On this basis the analysis of the interstitial/vacancy nature of $\frac{1}{3}\langle 111 \rangle$ loops can be carried out in a very simple way. Consider for example images taken of faulted loops with **B** $= [111]$ using the three 220 reflections. Loops which go out of contrast with **g** $= \bar{2}20$ have **b** $= \frac{1}{3}[\bar{1}\bar{1}1]$ or $\frac{1}{3}[11\bar{1}]$. Images taken with **g** $= 002$ with **B** $\sim [110]$ will give weak contrast if **b** $= \frac{1}{3}[\bar{1}\bar{1}1]$ since **g·b** will be positive and further for **B** $= [110]$ the upward normal to the loop plane would be $[11\bar{1}]$ and the loop would be vacancy if weak contrast was observed.

A similar analysis can be applied to faulted loops in other crystal systems if **g·b** is small, e.g. if **b** $= \frac{1}{2}[0001]$ and **g** $= \pm 10\bar{1}1$ weak contrast will be observed for **g** $= 10\bar{1}1$ since **g·b** will be positive.

5.3.11 Imaging of magnetic samples

The deflection of the incident electrons caused by a magnetic sample can be used in several different ways to reveal magnetic domains. The simplest case to consider is the contrast expected from 180° domain boundaries separating alternate domains with the magnetisation in the plane of the foil but in opposite directions [17]. If the electron distribution is considered just below the bottom of the thin foil it is apparent that alternate domain boundaries will have either an excess or a deficiency of electrons associated with them. Thus an out of focus or dark field image will reveal alternate 180° domain boundaries as dark and bright lines. The width of these lines will be a function both of the difference in strength of the domains and the degree of defocusing. The necessity of defocusing to observe domain boundary contrast is a disadvantage if other fine details such as dislocations are of interest and in-focus images can be obtained in two ways. Firstly by selecting only one part of the emergent direct (or diffracted) beam (see Chapter 4) with the objective aperture, so that alternate domains will appear dark or bright – this contrast will reverse if the other part of the beam is selected. Secondly, in crystalline samples, the domains will show up by diffraction contrast because the deflection of the electrons by the magnetic field within the sample changes the diffraction conditions.

The ability to obtain images of magnetic domains together with crystal defects allows any interaction between defects such as antiphase domain boundaries in ordered alloys and magnetic domains to be observed directly (e.g. [18]).

5.4 INFLUENCE OF ELECTRON OPTICAL CONDITIONS ON IMAGES IN TEM AND STEM

Imaging in TEM and STEM is equivalent if the electron optical conditions are appropriately adjusted. On this basis the theory developed for TEM images can be applied to STEM images, but this can be done only if the probe convergence in STEM ($2\alpha_s$) is identical to the objective aperture size in TEM ($2\beta_c$) and in addition the collection angle in STEM ($2\beta_s$) is identical to the illumination angle in TEM ($2\alpha_c$). If these conditions are met then the principle of reciprocity [1] shows that the images must be identical. In practice this equivalence is seldom achieved and it is important to appreciate the significance of the departures from reciprocity. Thus the value of $2\alpha_s$ is typically 10^{-2} radians (in order to get sufficient current into the small probe) which is somewhat larger than the usual size of the TEM objective aperture which is typically about 5×10^{-3} radians. The collection angle in STEM, $2\beta_s$ must be less than $2\theta_B$, if only the direct or a diffracted beam is to be collected, which sets an upper limit of about 10^{-2} radians and in order to maximize the signal this large value is commonly used whereas $2\alpha_c$ is typically around 5×10^{-4} radians. Thus the convergence in STEM ($2\alpha_s$) is within a factor of two of the objective aperture in TEM ($2\beta_c$), but the collection angle in STEM ($2\beta_s$) is more than ten times the beam divergence in TEM ($2\alpha_c$).

Calculations of contrast in TEM are normally carried out making the assumptions that there is no beam divergence (i.e. $2\alpha_c = 0$) and that the objective aperture size ($2\beta_c$) is of the order of 5×10^{-3} radians; this value is implied by the assumed value of anomalous absorption parameter [4] used in the image calculations. These calculations cannot be carried over directly into STEM because of the large value of $2\beta_s$ which is equivalent to a large beam divergence ($2\alpha_c$) in TEM. Calculations of images must therefore be done taking account of this finite beam convergence for larger angles than those appropriate for divergence in TEM [19]. The origin of the changes in image contrast which are observed when a large beam divergence is used can be seen by reference to equation (5.8). The range of angles encompassed within the beam divergence leads to a range of values of s_g and hence to a range of values of ξ_g. In bright field the image will tend to be dominated by the region of best transmission (i.e. the region just positive of the Bragg condition) and the resultant image appropriate to a given value of $2\beta_s$ can be obtained by summing intensities of calculated images obtained over the appropriate range of angles, i.e. the incident probe is decomposed into a set of beams of angle of incidence ranging from $-\beta_s$ to $+\beta_s$ either side of the mean angle and the image intensities calculated for each beam before adding intensities. Intuitively it is obvious that, as the range of $2\beta_s$ is increased, information will tend to blur out because of the different periodicities associated with the different values of ξ_g^{eff}. Thus bend contours, thickness fringes and oscillations in the images of dislocations will tend to be made weak the larger $2\beta_s$ is made. It therefore follows that if weak diffraction contrast effects are to be investigated TEM is to be preferred to STEM. If $2\beta_s$ must be reduced in order to increase the contrast this leads to a reduction in signal although a field emission gun improves the STEM imaging dramatically. In some cases it may be desirable to damp out bend contours, or thickness contours but in general TEM imaging is more useful in defect analysis than is STEM imaging.

On the basis of the above discussion it should be clear that the images of defects in STEM can be interpreted using the theories developed for TEM, provided that the influence of beam divergence is allowed for by adding images over the range of angles defined by $2\beta_s$. Thus the whole of the section dealing with defect analysis in TEM can be carried over into STEM.

5.5 INTERPRETATION OF HIGH RESOLUTION ELECTRON MICROSCOPY IMAGES

High resolution electron microscopy (HREM) is now being fairly extensively used because microscopes with point resolution of $\lesssim 0.3$ nm are available commercially. As pointed out in Chapter 3 HREM images are obtained by allowing several Bragg diffracted beams, together with inelastically and elastically scattered electrons which appear between these beams, through the objective aperture. Interpretation of HREM micrographs is straightforward only if some very specific conditions are fulfilled [20]. Thus only under the appropriate imaging conditions, when the crystal is oriented with a prominent zone axis along **B**, the electron beam direction, will the columns of atoms appear dark and the tunnels between the atoms appear bright. In order for this simple projection approximation to be true the crystal must be very thin (where very thin means around 80 Å at 200 kV for a medium atomic weight sample), the imaging must be carried out at a specific value of defocus (see below) and the objective aperture must allow through only those diffracted beams which correspond to distances within the point resolution of the microscope. Finally the incident beam should be precisely along the optical axis of the microscope and any crystal defects should also lie along **B**.

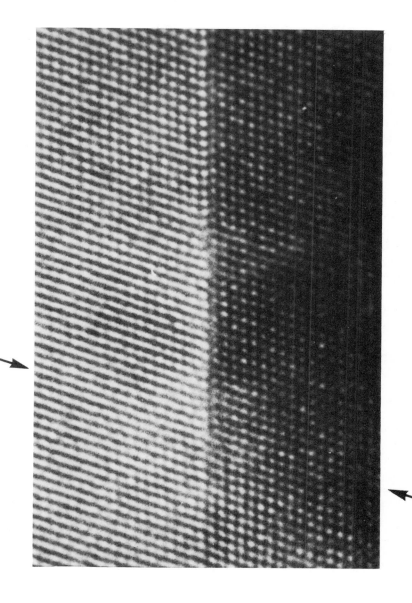

Fig. 5.14 High resolution electron micrograph taken at 200 kV from an (001) Si/NiSi$_2$ interface showing the apparent breakdown in epitaxy. Taken at Scherzer defocus (after reference 21).

If these conditions are not all fulfilled then the images can be safely interpreted only by carrying out detailed calculations. An example of a micrograph which illustrates the complications found with specimens, which exceed the very thin criterion, is shown in Fig. 5.14. The micrograph is of an (001) $Si/NiSi_2$ interface and appears to show that the epitaxy, evident at one part of the interface, is lost at other parts. In fact calculated images show that such an effect is expected in foils of changing thickness [21] and there is in fact complete epitaxy along the interface shown. The calculations consume much computer time since they are based on the multislice approach [22]. In such calculations the object is imagined to be divided into slices perpendicular to the incident beam and the intensity of scattered beams calculated as the electrons propagate through successive slices. If no change is observed in the calculated image when the thickness of the slice is reduced, the image is accepted.

The problems which arise in the interpretation of thicker samples and of images obtained at different values of defocus arise from the fact that the objective lens is imperfect and introduces phase shifts into high angle information. This problem is formally expressed by use of the contrast transfer function (CTF) of the objective lens. A schematic figure representing a CTF for an objective lens is shown in Fig. 5.15. Here the ordinate is a parameter representing the phase shift which can vary between 1 and -1 and the abscissa is in reciprocal space so that it is typically in units of nm^{-1}. This function shows that the phase is negative until a certain value of **g** after which very rapid oscillations are seen. If the defocus is changed so this CTF

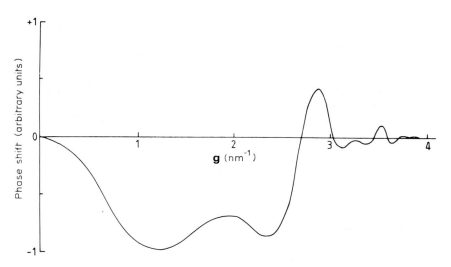

Fig. 5.15 Schematic contrast transfer function for a modern 100 kV object lens.

changes. It has been shown (e.g. [23]) that if the objective aperture acceptance angle corresponds to the first zero in the CTF then the image will have the correct phase relationships and will thus be interpretable. As indicated above, computing will be required if the thickness is greater than a specific value typically about 8 nm – given by about $\xi/2$, where ξ is the main extinction distance appropriate to the many beam imaging conditons. The appropriate value of defocus, which corresponds to a maximum value along the abscissa for the first zero, is given by $(C_s \lambda)^{1/2}$ (termed Sherzer focus) and $1.2\,(C_s\lambda)^{1/2}$ for optimum defocus.

The value of defocus is selected experimentally by finding precise focus (i.e. with no aperture, select the image showing minimum contrast) and then defocusing the objective lens the appropriate distance.

The actual value of the first zero is given by $0.6\ C_s^{1/4}\lambda^{3/4}$ so that the (reciprocal) value increases as C_s and λ get smaller as would be expected, i.e. the lens produces a more faithful image to larger values of **g**, corresponding to larger scattering angles, the smaller are C_s and λ.

Images obtained where the objective aperture size corresponds to the first zero in the transfer function at optimum defocus are thus interpretable immediately if $t \lesssim \xi^{main}/2$ and the resolution in such micrographs corresponds to the point resolution of the microscope. With currently available microscopes with a point resolution of the order of 0.3 nm this means that pairs of atoms in silicon are resolved but not individual atoms. It is possible however to obtain images which contain higher resolution information than implied by the C_s of the objective lens (the resolution being limited instead by the electronic instabilities in the microscope) by using a larger objective aperture. If images using the appropriate contrast transfer function are computed for assumed structures and are compared with the experimental images the best match (perhaps) corresponds to the real structure.

High resolution images can be obtained in STEM provided the same constraints as described above are met with. Additionally the convergence angle must be large enough for beam overlap to occur since this allows the necessary interference to occur between the direct and diffracted beams. Correct phase relationships will be maintained only if the illumination is coherent across the whole of the convergent probe.

5.6 INTERPRETATION OF SCANNING ELECTRON MICROSCOPY IMAGES

The interpretation of images in SEM is straightforward compared with the interpretation in STEM or TEM, because of the relative insensitivity to crystallographic influences on electron specimen interactions and to their overall similarity to optical micrographs; but it is nevertheless necessary to consider the factors which lead to, and which limit, the information in

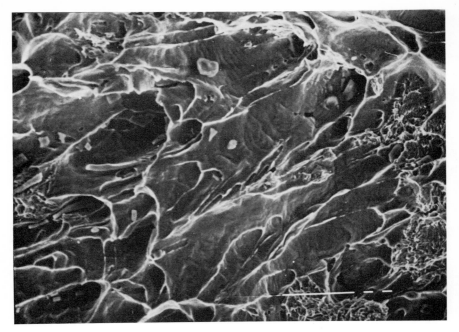

(a)

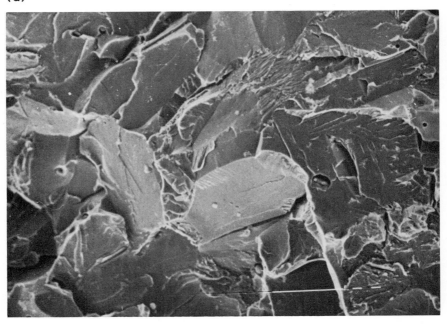

(b)

Fig. 5.16 Scanning electron micrographs using secondary electrons showing (a) ductile and (b) brittle fracture. Taken at 20 kV, illustrating the large depth of field obtainable in SEM. The bar is 10 μm. (After M.G. Hall.)

SEM images. In the case of magnetic samples contrast is observed from the individual domains and the origin of this contrast will be briefly considered below.

If only secondary electrons are collected (i.e. electrons with energies $\leqslant 50$ eV) the contrast can arise from the following sources: changes in surface topology, (since if the surface is locally nearly parallel with the electron beam direction a high secondary electron signal will be generated since a large number of incident electrons will be elastically scattered so that they travel near the suface of the sample), surface magnetic fields, surface electric fields, and differences in secondary electron emission characteristics of phases. Of these the topological contrast is the most widely used. Because the parts of the specimen which have the most direct path to the detector appear brightest, the image of a rough specimen appears as if it is viewed from above, when being illuminated from the detector. The fact that the low energy secondary electrons are bent into the detector from regions which do not have a direct line of sight into the detector allows details within holes and shaded regions to be observed – especially if standard image processing is used to enhance contrast within the dark regions. The strength of the SEM in the topological mode lies of course in the very large depth of field as discussed in Chapter 3 (cf. Fig. 5.16).

If second phase particles are present in the specimen it is possible that the intensity of the image will be influenced by the relative efficiency of secondary electron production and this may make interpretation of the topology less straightforward. Stereo pairs taken by tilting the specimen through the appropriate angle can remove any such problem and in addition make the interpretation of any height difference far more obvious. As in TEM, the tilt angle between stereos which are suitable for an individual can vary but are influenced by the height difference and the magnification. If there are large changes in height (which would require a small divergence angle to obtain an adequate depth of field) and if the magnification is high then the required tilt angle between the stereo pairs will be small. Typically 10° tilt is appropriate for a print magnification of 1000 times from a rough fracture surface. The stereo pair should be viewed in a stereoviewer with the tilt axis perpendicular to the line joining the viewer's eyes. Measurements of height difference can be made on most commercial stereoviewers provided the tilt angle and magnification are accurately known. It should be noted that the magnification is influenced by the tilt angle; the influence being largest in the direction perpendicular to the tilt axis. This effect can be compensated for on-line on modern SEMs and STEMs.

If backscattered high energy electrons are used to image the sample the contrast mechanisms, which underlie the common imaging modes, involve atomic number contrast and orientation contrast. Orientation contrast arises because, as discussed earlier, the backscattered intensity is a function of the incident angle of the electron beam with respect to the crystal planes. The intensity backscattered from a crystal for a given electron beam direction is thus determined by the orientation of that particular crystal and the intensity variation from a polycrystalline sample will cover the range in intensities visible in channelling patterns. This imaging mode, although one of low contrast is very useful when examining specimens which are difficult to etch or which, because of the particular experiment, cannot be etched. Again a minimum beam divergence will maximize the contrast but as is evident from Chapter 1 the image intensity decreases with decrease in divergence angle and signal: noise problems can reduce the contrast; only by increasing gun brightness and/or increasing the scan time can this problem be overcome. It cannot be emphasized too strongly that imaging in SEM requires the operator to select the appropriate balance between probe size, beam divergence and magnification in order to optimize conditions for each application.

Atomic number contrast arises because the backscattering efficiency is a function of atomic weight; heavy elements scatter electrons more efficiently

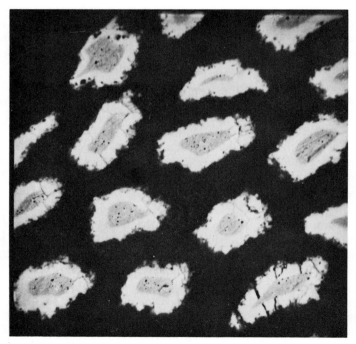

Fig. 5.17 Scanning electron micrograph using backscattered electrons (30 kV) showing atomic number contrast in Nb_3 Sn wires in a copper matrix. (After M.G. Hall.) Mag. × 2000.

than do light elements. It is therefore possible to use backscattered images to detect differences in composition if the difference results in a change in the mean atomic weight (cf. Fig. 5.17). The technique is in fact sensitive enough to detect differences corresponding to a change in mean atomic weight of less than one atomic mass unit.

The contrast observed from magnetic samples has been classified as Type I and Type II magnetic contrast. (For a summary of references see [24].) In Type I contrast the magnetic fields above the surface of the sample influence the trajectory of the low energy secondary electrons. Hence contrast between domains is observed when the secondary electron signal is used. In Type II contrast, which is observed with secondary, backscattered and specimen current signals, the trajectories of electrons within the samples are influenced by the Lorenze force due to the magnetization. With tilted specimens the effect is to influence the trajectories of electrons in particular domains so that they travel close to the surface. In these domains the backscattered and secondary electron yield will therefore be increased and contrast observed. These specialized applications of SEM imaging are summarized in [24] and are not elaborated further here.

REFERENCES

1. Cowley, J.M. (1979) *Introduction to Analytical Electron Microscopy*, (eds J.J Hren, J.J. Goldstein and D.C. Joy), Plenum, New York.
2. Read, W.T. (1953) *Dislocations in Crystals*, McGraw-Hill, New York.
3. Head, A.K., Clarebrough, L.M., Humble, P. and Morton, A.J. (1973) *Computed Electron Micrographs and Defect Identification*, North Holland, Amsterdam.
4. Hirsch, P.B., Howie, A., Nicholson, R.B., Pashley, D.W. and Whelan, M.J. (1965) *Electron Microscopy of Thin Crystals*, Butterworths, Sevenoaks.
5. Marukawa, K. (1979) *Phil. Mag.*, **40**, 303.
6. Maher, D.M. and Eyre, B.L. (1971) *Phil. Mag.*, **23**, 409.
7. Foll, H. and Wilkens, M. (1975) *Phys. Stat. Solidi.*, **A31**, 519.
8. Loretto, M.H. and Smallman, R.E. (1976) *Defect Analysis in Electron Microscopy*, Chapman and Hall, London.
9. Ishida, Y., Ishida, H., Kohra, K. and Ichinose, H. (1980) *Phil. Mag.* A, **42**, 453–62.
10. Rhule, M., Wilkens, M. and Haussermann, F. (1971) *Conf. on Microscopy of Cluster Nuclei, Chalk River*.
11. Ashby, M.F. and Brown, L.M. (1963) *Phil. Mag.*, **8**, 1083.
12. English, C.A., Eyre, B L. and Holmes, S.M. (1980) *J. Phys. F.: Metal Phys.*, **10**, 1065.
13. Matsumara, S., Toyoharo, M., Tomokiyo, Y., Oki, K. and Eguchi, T. (1982) *Proc. 10th Int. Congr. on Electron Microscopy, Hamburg*, Vol. 2, Deutsche Gesell. Elektromikroscopie, p. 87.
14. Haque, E., Jones, I.P. and Loretto, M.H. (1975) *Developments in Electron Microscopy and Analysis* (EMAG 75), (ed. J.A. Verables), Academic Press, London, p. 429.
15. Brooks, J.W., Loretto, M.H. and Smallman, R.E. (1979) *Acta Met.*, **27**, 1839.
16. Chou, C.T. and Hirsch, P.B. (1981) *Phil. Mag.*, **44**, 1415.

17. Hale, M.E., Fuller, H.W. and Rubinstein, H. (1959) *J. Appl. Phys.*, **30**, 781.
18. Lapworth, A.J. and Jakubovics, J.P. (1974) *Phil. Mag.*, **29**, 253.
19. Whelan, M.J. and Hirsch, P.B. (1957) *Phil. Mag.*, **2**, 1121.
20. Cowley, J.M. (1978) *J. Nucl. Mater.*, **69**, 228.
21. Cherns, D., Spence, J.C.H., Anstis, G.R. and Hutchison, J.L. (1982) *Phil. Mag.*, **A46**, 849.
22. Moodie, A.F. (1972) *Z. Naturfosch.*, **27**, 437.
23. Spence, J.C.H. (1981) *Experimental High Resolution Electron Microscopy*, Clarendon Press, Oxford.
24. Wells, O.C. (1982) *Microbeam Analysis*, San Francisco Press, p. 447.
25. Jones, I.P. and Hutchinson, W.B. (1981) *Acta Metall.*, **29**, 951.

6
INTERPRETATION OF ANALYTICAL DATA

6.1 INTERPRETATION OF X-RAY DATA

Qualitative interpretation of X-ray data simply requires that the energy (or wavelength) of the observed peaks be compared with the energies of the characteristic peaks of the elements. The quantitative interpretation of X-ray data from thin samples differs from the interpretation of similar data from thick samples in several ways. The whole of the following discussion of the interpretation of X-ray data will be based on the assumption that the data have been obtained from a thin film using an EDX rather than a WDX system. Clearly the underlying principles are identical to those for the interpretation of bulk analysis using EDX or WDX detectors. The interpretation of data from bulk samples is more involved and the major differences between thin foil and bulk analysis are discussed in Section 6.3 and have also been covered in Chapter 2.

6.2 INTERPRETATION OF DATA FROM THIN SAMPLES

6.2.1 X-ray data from thin samples

All quantitative X-ray microanalysis requires that the background bremsstrahlung be subtracted from the observed spectrum so that the relative intensities of the characteristic X-rays can be determined. As discussed in Chapter 2 the general form of the background X-ray intensity can be modelled and expressions such as that in equation (2.25) can be used, in principle, in conjunction with expressions which allow for absorption in the specimen and in the detector window, to calculate the expected shape of the continuum radiation. Especially at the low energy end this is not a practical proposition (since absorption is then a sensitive function of composition and the composition is unknown) and it is common practice therefore to fit an empirical background visually on the spectrum when it is displayed on the visual display unit (VDU). With flexible programming it is possible to observe and modify a fitted background which appears to model accurately the experimental background (cf. Fig. 6.1). This empirical approach is satis-

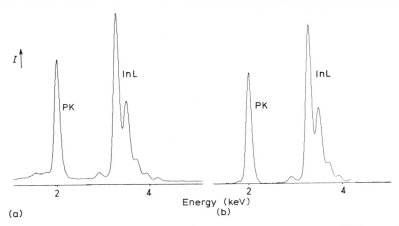

Fig. 6.1 Typical EDX output from a thin indium phosphide sample (a) before and (b) after stripping an empirically fitted background.

factory provided there are regions in the background which are clear of characteristic peaks so that the variation of background intensity with energy is obvious. There are occasions when examining complex alloys where peaks are situated very close to each other and the empirical approach to background stripping, outlined above, cannot be used. In such cases it may be possible to use either standard pure metal or simple alloy spectra to subtract out the background and some of the peaks, but because the details of any approach tend to be specific to each case (or in some cases specific to the available software) it is not useful to discuss the situation any further.

There are several simplifications which arise from the use of thin foils, where thin is used in this context to describe samples used in transmission electron microscopy. The most important of these arises from the fact that the average energy loss which electrons suffer on passing through a thin foil is only about 2%, and this small average loss means that the ionization cross section (equation (2.14)) can be taken as a constant. Thus the number of characteristic X-rays photons generated from a thin sample is given simply by the product of the electron path length and the appropriate cross section Q, and the fluorescent yield ω. Both Q and ω are defined in Chapter 2. On this basis it can be seen immediately that the intensity generated by element A is given by

$$I_A = iQ\omega n \qquad (6.1)$$

where Q is the cross section for the particular ionization event, ω the fluorescent yield, n the number of atoms in the excited volume, and i the current incident on the specimen. Microanalysis is usually carried out under conditions where the current is unknown and interpretation of the analysis simply requires that the ratio of the X-ray intensities from the various elements be obtained. If specimens are being analysed at different times, say if standards are being used, it may be necessary that the current, whilst being unknown absolutely, remains constant.

(a) Thin film approximation

If we consider the simple case of a very thin specimen of unit thickness for which absorption and X-ray fluoresence can be neglected, then the measured X-ray intensity from element A generated by a current i is given by

$$I_A = n_A Q_A \omega_A a_A \eta_A i \tag{6.2}$$

and for element B by

$$I_B = n_B Q_B \omega_B a_B \eta_B i \tag{6.3}$$

where a and η represent the fraction of the K line (or L and M) which is collected and the detector efficiencies respectively for elements A and B.

Thus in an alloy made up of elements A and B

$$\frac{n_A}{n_B} = \frac{I_A Q_B \omega_B a_B \eta_B}{I_B Q_A \omega_A a_A \eta_A} \tag{6.4}$$

which is generally written

$$\frac{n_A}{n_B} = k_{AB} \frac{I_A}{I_B} \tag{6.5}$$

and this equation forms the basis for X-ray microanalysis of thin foils (e.g. [1], [2]), where the constant k_{AB} contains all the factors needed to correct for atomic number differences, i.e. the Z-correction.

On the assumption that the background bremsstrahlung has been removed and that the data are in the form of the measured intensities at full width half maximum (FWHM), expression (6.4) can be used to calculate the ratio of the number of atoms of A to the number of atoms B, i.e. the concentrations of A and B in the binary alloy.

The equations which allow Q and ω to be calculated have been discussed in Chapter 2 and the detector efficiency in Chapter 3. These data, taken together with the appropriate value of a (see equation (6.3)), mean that it is straightforward to obtain C_A/C_B from the measured intensities. An example is shown in Table 6.1 where the values of I_K^{Al} and I_K^{Ni}, obtained after stripping the background, are shown in the first column.

In view of uncertainties in the effective thickness of the window on the detector (which may change if contamination builds up), and in view of the uncertainties in values for Q and ω, it is advisable to check the analytical system by using a number of standard alloys or pure metals of known thickness. It has been common practice to use mineral standards of known

Table 6.1 The relationship between measured intensities and composition for a NiAl alloy.

Measured (FWHM) intensities		Cross section $Q(\times 10^{-24} \text{cm}^2)$	Fluorescent yield, ω	$\dfrac{K_\alpha}{K_\alpha + K_\beta}$	Detector efficiency, η	Analysis (at.%)
Ni	K_α	18 851	297	0.392	0.88	0.985	54.3
Al	K_α	8 327	2935	0.026	0.98	0.725	45.7

composition (or some other very thin standards) to normalize all values of the constant in equation (6.5) to silicon so that k_{ASi} for the different elements was obtained [1]. A typical curve for K lines is shown in Fig. 6.2, which of course has $k = 1$ for Si. These curves effectively represent sensitivity factors incorporating both generation and detection efficiencies (if absorption and fluorescence are negligible) for all the elements and hence allow instant calculation of composition from relative intensities. These k curves for K lines (and corresponding curves for L and M lines) are very useful but of course should be used only on the particular analytical instrument on which they are obtained. With dedicated computers on analytical instruments it is common practice to use equation (6.4) to compute C_A/C_B where C_A and C_B are the (initially unknown) concentrations of elements A and B in the AB alloy, rather than use curves such as that in Fig. 6.2.

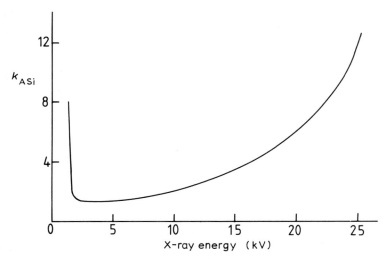

Fig. 6.2 Experimental values of the constant k_{ASi} at 100 kV for the K lines (after [1]). See text.

The absolute accuracy of any X-ray analysis depends either on the standards used or on the accuracy of the constants in equation (6.4). Especially in the case of L or M lines the constants are not known with any confidence and it is highly desirable to use standards to calibrate analyses. If standards are used it is essential to realize that the accuracy of all subsequent analyses is determined by the accuracy to which the standard analysis has been carried out. The generation of X-rays is subject to normal statistical uncertainties and the accuracy depends on the number of X-ray photons collected. Thus for a 95% confidence limit the percentage accuracy is given by $(100 \times 2\sqrt{N})/N$, where N is the number of photons counted. For example for an accuracy of 2% (with a 95% confidence limit) N should be 10^4 and for a 0.2% accuracy N should be 10^6. Because of the limited count rate of EDX detectors (see Chapter 3) very long counting times should be used when calibrating using standards in order to ensure that the obtainable accuracy of future analyses is not limited.

(b) Absorption and fluorescence corrections for thicker foils

In all samples there is a finite possibility, either that some characteristic X-rays will be absorbed more than others, or that the intensity of one of the characteristic X-ray lines will be increased by fluorescence. Absorption of one X-ray with respect to another will be more likely the further apart are the X-ray energies; fluorescence is most likely to occur when two elements are two apart in the periodic table for $Z > 21$ and one apart for $Z < 21$. From the point of view of quantitative microanalysis of thin foils the importance of absorption and fluorescence must be assessed to see if they are significant in terms of the statistical accuracy of the measurement. Thus, if the measurement of the intensities of the X-rays has been carried out with a statistical error of say 10% then there is little point in carrying out a correction for fluorescence/absorption if the maximum influence is only 2%. The method of assessing the significance of absorption and fluorescence and for applying necessary corrections will be dealt with in the following sections.

(i) Absorption correction

The attenuation of X-ray photons of an incident beam of intensity I_0 in a path length l in a material of density ρ is given by the relation

$$I = I_0 \exp\left[-\left(\frac{\mu}{\rho}\right)\rho l\right] \tag{6.6}$$

where μ/ρ is the mass absorption coefficient of the X-ray in the specimen (in units of $cm^2 g^{-1}$). The way in which the mass absorption coefficient varies with X-ray energy for a particular sample is shown in Fig. 6.3. The step in the function at X corresponds to an X-ray energy of the appropriate

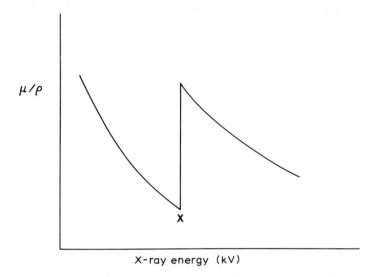

Fig. 6.3 Curve showing the energy dependence of the mass absorption coefficient for X-rays. The jump at X corresponds to an absorption edge where the incident X-rays have just sufficient energy to eject an inner shell electron.

value to displace an electron in the sample, i.e. this is an ionization edge and at this value the absorption coefficient increases typically by a factor of between 2 and 20.

Since in thin foils the probability of generating an X-ray can be taken as constant through the sample (the mean energy loss is only ~2%) the ionization cross section for X-ray production, Q, can be taken as constant and the correction for absorption can be carried out by integrating through the foil for the appropriate geometry. This requires a knowledge of the specimen thickness and a knowledge of the X-ray take-off angle and specimen tilt. The determination of the specimen thickness* can be done using any of the techniques discussed in Chapters 4 and 5 or by the plasmon technique (Section 6.3).

The detector–specimen geometry is typically that shown in Fig. 6.4 where the angle of elevation of the detector is θ_E and the tilt angle of the specimen towards the detector is θ_T. Then the X-ray path length l is given by

$$l = \frac{t \sin \theta_T}{\cos (\theta_T - \theta_E)} \tag{6.7}$$

* It should be noted that the ratios of the thickness of areas of a homogeneous sample can be obtained directly from the ratios of either the integrated bremsstrahlung or the characteristic X-ray intensity for a constant electron dose. Thus the absolute thickness needs to be determined at only one point on a sample and the thickness of other regions obtained by simple proportion.

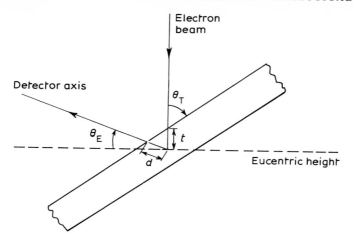

Fig. 6.4 Typical geometry for EDX detector in a transmission electron microscope. The angle of elevation of the detector is θ_E and the tilt angle of the specimen towards the detector is θ_T.

where t is the depth in the foil at which the X-rays are generated corresponding to the angle θ_E; since θ_E is taken to be a constant this is tantamount to assuming that the X-rays are generated at the foil centre.

The intensity of X-rays corrected for absorption, I_A^{abs}, exiting from the sample containing elements A and B is then given by

$$I_A^{abs} = \int_0^{t_0} I_A \exp\left[-\left(\frac{\mu}{\rho}\right)_{AB}^A \rho_{AB} \frac{t \sin\theta_T}{\cos(\theta_T - \theta_E)}\right] dt \qquad (6.8)$$

where $(\mu/\rho)_{AB}^A$ is the mass absorption coefficient for the X-ray from A in the specimen AB, I_A is the (constant) generation rate of X-rays though the sample given by the product of i, Q and ω (equation (6.1)), and t_0 is the sample thickness in the electron beam direction. Integration of this leads to

$$I_A^{abs} = I_A \left\{1 - \exp\left[-\left(\frac{\mu}{\rho}\right)_{AB}^A \rho_{AB} t_0 \frac{\sin\theta_T}{\cos(\theta_T - \theta_E)}\right]\right\} \left[\left(\frac{\mu}{\rho}\right)_{AB}^A \rho \frac{\sin\theta_T}{\cos(\theta_T - \theta_E)}\right]^{-1} \qquad (6.9)$$

A similar expression can be written for I_B^{abs}, thus

$$I_B^{abs} = I_B \left\{1 - \exp\left[-\left(\frac{\mu}{\rho}\right)_{AB}^B \rho_{AB} t_0 \frac{\sin\theta_T}{\cos(\theta_T - \theta_E)}\right]\right\} \left[\left(\frac{\mu}{\rho}\right)_{AB}^B \rho \frac{\sin\theta_T}{\cos(\theta_T - \theta_E)}\right]^{-1} \qquad (6.10)$$

so that

$$\frac{I_A^{abs}}{I_B^{abs}} = \frac{I_A X_B}{I_B X_A} \frac{1 - \exp(-X_A \rho_{AB} t_0)}{1 - \exp(-X_B \rho_{AB} t_0)} \quad (6.11)$$

where

$$X_A = \left(\frac{\mu}{\rho}\right)_{AB}^A \frac{\sin \theta_T}{\cos(\theta_T - \theta_E)} \quad (6.12)$$

and

$$\left(\frac{\mu}{\rho}\right)_{AB}^A = \left(\frac{\mu}{\rho}\right)_A^A C_A + \left(\frac{\mu}{\rho}\right)_B^A C_B \quad (6.13)$$

Equation (6.13) makes it clear that the mass absorption coefficient in alloy AB for X-ray A depends on the weighted values of the mass absorption coefficients in pure A and pure B. Similar expressions to (6.12) and (6.13) can thus be written for element B.

Substitution and rearrangement of equation (6.11) leads to the following expression for calculating the fractional composition of the AB alloy:

$$\frac{C_A}{C_B} = \frac{I_A^{abs}}{I_B^{abs}} = \frac{I_A}{I_B} \frac{Q_B \omega_B \eta_B a_B A_A X_A}{Q_A \omega_A \eta_A a_A A_B X_B} \frac{1 - \exp(-X_B \rho_{AB} t_0)}{1 - \exp(-X_A \rho_{AB} t_0)} \quad (6.14)$$

Thus if the values of I_A^{abs}, I_B^{abs} are obtained experimentally and values of Q_i (the ionization cross section), ω_i (the fluorescent yield), η_i (the detector efficiency), a_i (the $K_\alpha/(K_\alpha + K_\beta)$ fraction), ρ_{AB} (the density of the alloy), $(\mu/\rho)_{AB}^i$ (the mass absorption coefficients and t_0 (the thickness along **B** the beam direction) are all known, C_A/C_B can be calculated. With dedicated computers this equation can be solved (on-line if t_0 is known) routinely.

In order to ascertain whether an absorption correction is worthwhile, in terms of the accuracy required and the statistical accuracy of the analysis, various suggestions have been made (e.g. [3], [4]). As is intuitively evident, absorption is significant if the values of $(\mu/\rho)_{AB}^A$ and $(\mu/\rho)_{AB}^B$ are very different. However, the calculations required to assess the significance of absorption, whilst being instructive, are only a little easier to carry out than the actual correction since a value of t_0 is required and computer solution of equation (6.14) may just as well be calculated. Of course this equation must be solved iteratively since the absorption correction is a function of the composition (equation (6.13)). For the first iteration the composition obtained from equation (6.5) is used.

(ii) *Fluorescence correction*

Fluorescence of a characteristic X-ray can occur if the energy of the characteristic X-ray from one of the components* is significantly higher than

---

* The bremsstrahlung intensity in thin foils is too low to give rise to significant fluorescence.

the energy of the other (see Fig. 6.3). The correction is negligible for all elements for $Z < 20$ because the cross section is negligible, and is a possible complicating factor only for the transition elements and then only if two such elements of Z and $(Z + 2)$ exist in the alloy.

In a specimen, such as Fe/Cr, where fluorescence may be significant the total intensity from element A, I_A^{tot} is given by an expression of the form

$$I_A^{tot} = I_A^0 + I_A^{FL} \qquad (6.15)$$

where I_A^0 and I_A^{FL} are the X-rays generated by the incident electrons and by fluorescence respectively. If we consider the case where there is only one element causing fluorescence the intensity of these X-rays I_B^0 will give rise to an intensity I_A^{FL} given by the appropriate cross section. Values for X-ray fluorescence cross sections are found from the values of mass absorption coefficients (e.g. [5]). The jump height across an absorption edge (cf. Fig. 6.3) is a measure of the increased probability of fluorescence at that X-ray energy. The mass absorption coefficient on the low energy side is associated with lower energy ionization events and the increase at the K absorption edge is due solely to K shell ionization. Thus if the ratio of the values of the mass absorption coefficients either side of the absorption edge is R_A^{abs} the fraction

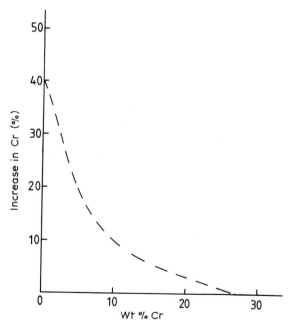

Fig. 6.5 Curve showing the composition dependence of the fluorescence correction for an Fe/Cr alloy (after [8]).

of all ionization events which give rise to X-rays from element A is given by

$$F = \frac{R_A^{abs} - 1}{R_A^{abs}} \qquad (6.16)$$

The derivation of the equations for fluorescence correction is very cumbersome and the reader is referred to [6] and [7]. The equation for fluorescence corrections for the fluorescence of K_α radiation of element A by the K X-ray of element B derived in these references is

$$\frac{I_A^{FL}}{I_A} = \omega_B C_B \frac{r_A^K - 1}{r_A^K} \left(\frac{\mu}{\rho}\right)_A^{BK_\alpha} \frac{U_B \ln U_B}{U_A \ln U_A} \frac{\rho t}{2} \left\{ 0.923 - \ln\left[\left(\frac{\mu}{\rho}\right)_{AB}^{BK_\alpha} \rho t\right]\right\} \qquad (6.17)$$

where $U_B = E_0/E_K^B$. The most often quoted example of a fluorescence correction is that of Cr in Fe; in this case the correction can be large and errors of a factor of two may be incurred if the correction is not carried out. The correction is a sensitive function of composition (and of specimen thickness) as shown in Fig. 6.5, for a specimen about 200 nm thick.

6.2.2 Influence of diffraction conditions on X-ray analysis

As discussed in Chapter 5 electrons are strongly absorbed by crystals when $s_g < 0$, i.e. when the crystal is set with the angle of incidence of the electrons just less than the Bragg angle. Both theoretical and experimental work has shown that this increased absorption is associated with enhanced X-ray production. Theoretically a very large (factor of ~ 2–3) increase in X-rays is predicted (e.g. [9]) but experimentally only a relatively small effect is observed in STEM; about a 15% increase. The relatively small increase occurs because the initial probe is highly convergent in STEM and therefore only a small fraction of the incident electrons are at angles which correspond to the conditions $s_g < 0$ even when the average beam direction is just below θ_B. Nevertheless it is clearly essential when carrying out either EDX or EELS analysis to ensure that the mean beam direction corresponds to the condition $s_g \gg 0$ for low order reflections. It is preferable to check that this condition is satisfied by examining the diffraction pattern rather than the image (see Section 2.3.3).

6.3 INTERPRETATION OF X-RAY DATA FROM BULK SAMPLES

In contrast to the situation for thin samples, electrons are either backscattered, usually with some earlier energy loss, or lose all their energy within the sample. The modelling of backscattering and the change of ionization cross section with path length has been outlined in Chapter 2.

In addition to these complications it is clear that absorption of X-rays and X-ray fluorescence will be far more significant in bulk samples than in thin foils.

The calculation of absorption and fluorescence corrections from first principles is not possible and all solutions used are, to some extent, empirical; a considerable amount of work has been carried out aimed at obtaining data which can be used to check the empirical equations. Absorption corrections can be checked by measuring the reduction in intensity of characteristic X-rays from element A by using various known thicknesses of element B to absorb the X-rays (e.g. [10]). Although these experiments are useful in checking the equations which have been developed, most quantitative analysis of bulk samples is done either by reference to standards or by using semi-empirical computer programs. Because analysis is virtually always done by these technique – standards plus computer based analysis – and because the equations which are used are very complex the reader is referred to specialist texts for further information (e.g. [11], Chapter 14 and 15).

6.4 INTERPRETATION OF ELECTRON ENERGY LOSS SPECTRA

A typical spectrum produced by EELS is shown in Fig. 6.6 and it can be seen that there are three distinct regions. The dominant feature on the left of the figure is termed the zero loss peak. This is made up firstly by those electrons which have not been scattered by the specimen, secondly by those that have suffered phonon scattering ($\sim 1/40$ eV) and thirdly by elastically scattered electrons. The angular range of scattering and the cross section for these scattering processes have been defined in Chapter 2. The energy width of the zero loss peak is caused by the energy spread of the electron source (up to ~ 2 eV for a thermionic tungsten filament) and the energy resolution of the spectrometer (typically ~ 1 eV).

The second region of the typical EEL spectrum shown in Fig. 6.6 contains electrons which have excited plasmons and extends up to about 50 eV loss. The intensity in this region is comparable with that in the zero loss peak. As pointed out in Chapter 2 the energy loss associated with plasmon excitations is 15–20 eV and this region of the energy loss spectrum commonly shows a series of equally spaced peaks of decreasing height corresponding to electrons which have suffered one, two, or more plasmon interactions. Since the typical mean free path for the generation of a plasmon is about 50 nm, many electrons suffer single plasmon losses but only in specimens which are far too thick for electron energy loss analysis will there be a significant third plasmon peak.

There are two significant applications of plasmon loss peaks in microanalysis. Firstly the shift in the energy of the peak can be used to measure

164   ELECTRON BEAM ANALYSIS OF MATERIALS

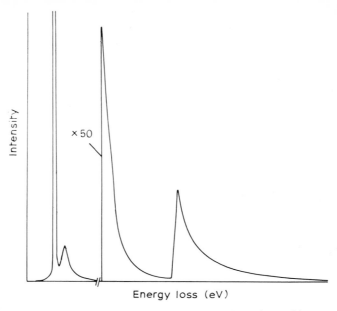

Fig. 6.6 Typical EELS spectrum showing the three main regions of interest: the zero loss, loss up to 50 eV, and the ionization losses. The intensity scale is × 50 for energies above 100 eV.

composition changes in some alloys. The number of alloys for which there is a significant shift in energy is limited and this together with the poor spatial resolution [12], makes this technique less important for microanalysis than measurement of ionization losses.

Secondly the relative size of the plasmon loss peak and the zero loss peak can be used to measure the foil thickness. Thus the probability P_n of an electron exciting (n) plasmons, and losing $n\, E_p$ is given by

$$P_n = \frac{1}{n!} \left[\frac{t}{L_p} \right]^n \exp\left(-\frac{t}{L_p} \right) \qquad (6.18)$$

where t is the thickness, L_p the mean free path for a plasmon excitation of energy E_p [12]. The value of t/L_p is given by the ratio of the probabilities of exciting a single plasmon to not exciting a plasmon, i.e. P_1 to P_0 and is given by the ratio of the intensities of the first plasmon peak to the zero loss peak. It should be noted, that if the second plasmon peak is a significant fraction of the first peak the specimen will be too thick for accurate analysis using the ionization losses which make up the third region of the EEL spectrum.

The third region in Fig. 6.6 is typically scaled by a factor of 50 or more relative to the first two regions and it can be seen to consist of a continuous background on which the characteristic ionization losses are superimposed. Qualitative analysis is carried out simply by measuring the energy of the edges and comparing them with tabulated energies. It is useful to compare the actual shapes of the edges with those available in the literature [13], since this can add confidence to the identification and can also help to define the chemical state of the element. This latter aspect is discussed briefly later in this chapter.

6.4.1 Quantification of characteristic energy losses

Measurement of the ratios of the intensities of the electrons which have suffered ionization losses from elements A and B allows, in principle, the ratio of the numbers of A atoms N_A and B atoms N_B to be obtained simply from the appropriate ionization cross sections Q_K. Thus the number of A atoms will be given by

$$N_A = \frac{1}{Q_K^A}\left(\frac{I_K^A}{I_0}\right) \qquad (6.19)$$

and the number of B atoms by a similar expression so that

$$\frac{N_A}{N_B} = \frac{I_K^A Q_K^B}{I_K^B Q_K^A} \qquad (6.20)$$

where I_K^A is the measured intensity of the K edge for element A and similarly for I_K^B. The term I_0 is the measured intensity of the zero loss peak. This equation is apparently similar to equation (6.5) which applies to thin samples using EDX.

There are several factors which are important in obtaining I_K which in turn define the appropriate value of Q_K. The first obvious factor, which has to be assessed before obtaining I_K, is the removal of the background, so that only loss electrons remain. This background is due to the low energy tails associated with lower energy inner shell losses, plasmon losses, and valence electron excitations. It is not possible to model this complex background accurately and instead an empirical fit of the form AE^{-r} is used to fit the smoothly falling background for 50–100 eV on the high energy (lower loss) side of the edge. This background is then extrapolated below the edge (see Fig. 6.7) and the intensity above this background is taken as I_K. The second factor which is obvious from Figs 6.6 and 6.7, is that the inner shell intensity decays only very slowly on the low energy (i.e. higher loss) side of the edge. To obtain I_K (total) would therefore require extrapolation of the background through a large energy range ($\gtrsim 200$ eV) but because other edges may be present and because of the limited accuracy of the background

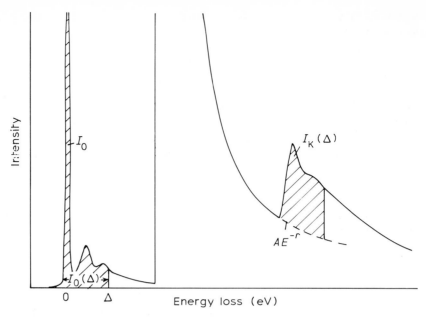

Fig. 6.7 Schematic energy loss spectrum showing the relation between the original spectrum, and the intensity I_K^A remaining after stripping the background up to an energy Δ above E.

fit this extrapolation is inaccurate*. Clearly therefore there is a maximum energy range over which I_K can be measured, which should not be less than 50 eV. Q_K in equations (6.19) must then be replaced by $Q_K(\Delta)$ which is a cross section calculated for atomic transitions within an energy range Δ of the ionization threshold.

The third factor, which must then be considered when obtaining I_K, arises because some electrons suffer double scattering events, which are in fact commonly visible as subsidiary peaks after the ionization edge (cf. Fig. 6.7). The integrated intensity $I_K(\Delta)$ therefore overestimates the true intensity of loss electrons and it is necessary to attempt to deconvolute the scattering events so that the true intensity of loss electrons over the energy range Δ may be obtained [14], [15]. A more simple approach which is virtually always used is to rewrite expression (6.19) as

$$N_A = \frac{1}{Q_K^A(\Delta)} \left(\frac{I_K^A(\Delta)}{I_0(\Delta)} \right) \tag{6.21}$$

---

* In some cases with closely spaced absorption edges it is not possible to fit the background and quantitative analysis cannot then be carried out without complex deconvolution.

where $I_0(\Delta)$ is the zero loss peak intensity measured now up to the energy loss Δ (see Fig. 6.7).

On the above basis $I_K^A(\Delta)$ can be obtained and related to N_A, via the cross section $Q_K^A(\Delta)$, but a final complication must be considered since it has been implicitly assumed that whatever angle the electrons are scattered through they will be collected. This is not true, although inelastic scattering is peaked in the forward direction, and a large fraction of the loss electrons are collected for practical values of scattering angle (see Chapter 3). Thus the scattering angle is $\Delta E/2E_0$ and if the scattering angle, measured at the specimen, is about 10 mrad then virtually all loss electrons up to 1 kV loss would be collected for $E_0 = 100$ kV. Typically about 50% of the loss electrons will be collected.

In practice it is therefore necessary to modify equation (6.21) further by use of a double partial cross section $Q(\Delta, \alpha)$, where α is the angular range of scatter at the specimen which is accepted by the spectrometer (see Fig. 3.8), so that

$$N_A = \frac{1}{Q(\Delta, \alpha)} \left(\frac{I_K^A(\Delta, \alpha)}{I_0(\Delta, \alpha)} \right) \tag{6.22}$$

where $I_K^A(\Delta, \alpha)$ is the area under the edge after stripping (but without deconvolution) up to energy Δ above the edge and similarly $I_0(\Delta, \alpha)$ is extended up to Δ as in equation (6.21). Thus analysis of a binary alloy is carried out using the following equation:

$$\frac{N_A}{N_B} = \frac{Q_K^B(\Delta^B, \alpha^B)}{Q_K^A(\Delta^A, \alpha^A)} \frac{I_K^A(\Delta^A, \alpha^A)}{I_K^B(\Delta^B, \alpha^B)} \tag{6.23}$$

The values of $Q(\Delta, \alpha)$ required for equation (6.22) for the specific value of ionization edge, Δ, α and incident accelerating voltage can be calculated by an approximate method for K and L lines [16] or by a longer, more exact method [17].

Virtually all analysis using EELS is done in this approximate way, without deconvolution, and it is clear that if accurate answers are required standards would be essential. The accuracy of analyses carried out using equation (6.22) would be no better than 10% even for a very thin sample and it is evident that the thicker the sample the less accurate the analysis since the ionization edge is convoluted with an increasing number of other energy loss events as the thickness increases, and this is corrected for only approximately.

The compositions derived from equation (6.23) may need further correction if the EELs spectrum has been collected using a highly convergent electron probe. Thus it has been shown [18] that the convolution of the incident beam convergence angle α_c with the spectrometer acceptance angle β

Table 6.2 Quantitative analysis of boron nitride using EELS and equation (6.22). Collection angle 10 mrad; partial cross sections calculated from [16].

	Intensity	Energy range, Δ (eV)	Partial cross section (b)	Composition (at. %)
B	538 814	71	12545	46
N	114 052	71	2260	54

reduces the collection efficiency by a factor R, given approximately by*

$$R = \frac{\ln\left[(1 + (\beta/\alpha_c)^2](\alpha_c/\theta_E)^2\right.}{\ln\left[1 + (\beta/\theta_E)^2\right]} \left(\frac{\beta}{\alpha_c}\right)^2 \quad (6.24)$$

where θ_E, the characteristic scattering angle, is given by $\theta_E = (E_c/2E_0)$, E_c being the energy of the loss electrons and E_0 the incident voltage. The significance of this can be appreciated by considering a 100 kV spectrum collected with $\beta = 4.5$ mrad from a sample containing titanium. β/θ_E would be 1 and equation (6.24) then shows that if $\alpha_c/\beta > 1$ there will be a very significant reduction in the intensity of the titanium signal. For example if $\alpha_c/\beta = 3$ (corresponding to a convergence angle of ~ 13 mrad), R will be about 0.4. Clearly spectra recorded in the STEM or probe mode will be influenced more than spectra recorded in the TEM mode.

Although the above analysis sounds complex, because of the problem in relating I_K^A and Q_K^A, the experimental procedure is relatively straightforward, since background stripping over defined energy ranges followed by integration of the remaining intensity, is done very rapidly. Calculation of N_A and N_B using equation (6.22) is then carried out using appropriately calculated values of $Q_K^A(\Delta, \alpha)$.

The following example serves to illustrate the technique and again as discussed in Section 5.12 it should be emphasized that all EELS and EDX analysis should be carried out with $\sim s_g \gg 0$ for all reflections to avoid enhanced ionization events. The intensities for boron and nitrogen are those intensities obtained after removing the background over the energy range specified in Table 6.3. The analysis is clearly in error by 4%.

In fact the information in ionization edges can sometimes provide additional information about the specimen if the sample is thin enough [16]. This is illustrated schematically in Fig. 6.7 where there is structure in the EELS spectrum both before and after the edge. The structure before the ioni-

---

* This expression assumes that the spectrometer acceptance angle β equals the scattering angle α, at the specimen. In the case of spectrometers which are efficiently coupled to the microscope (see Chapter 3), where α is typically between two and five times β, the influence of α_c is correspondingly reduced.

zation edge is a function of the density of unoccupied bound states in the sample and is influenced by the chemical and crystallographic states of the atom. The extended fine structure after the edge (which is analogous to the X-ray fine structure in X-ray absorption experiments (EXAFS)) extends for many hundreds of electronvolts from the edge. This structure arises because the loss electrons interact with the neighbouring atoms and electrons; the oscillatory behaviour in the spectrum is caused by the associated diffraction effect and gives information about local order around the excited atom.

6.5 INTERPRETATION OF AUGER SPECTRA

As pointed out in Chapter 3 the output from Auger spectrometers is virtually always presented as a differentiated spectrum over the appropriate

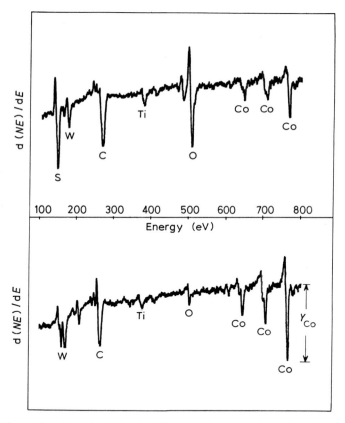

Fig. 6.8 Figure showing a typical output from an Auger spectrometer where the results are displayed as $d(NE)/dE$ against E. Taken from tungsten carbide (a) before and (b) after ion beam cleaning.

range of energy. This is done to increase the visibility of the Auger signals and to remove the continuously sloping background. A typical Auger output is shown in Fig. 6.8 and this type of output will be considered in this section.

The interpretation of Auger spectra can be carried out qualitatively, if the only information required is the identification of the elements present, or quantitatively if the ratios of those elements present are required. Qualitative analysis is carried out simply by comparing the observed energies with those in standard spectra obtained from pure standards. Auger electrons are produced with sharply defined energies (see Chapter 2) and although there may be overlap problems occasionally, especially with transition metals, there are usually peaks which do not overlap, which allow ambiguities to be removed. The major peaks should be identified first and this will reduce the number of possible elements responsible for any of the major peaks to two or perhaps three. Just as with electron energy loss it is then best to compare the observed peaks with a library of standard spectra [19] and to remove any ambiguities in this way. Again, as with energy loss, it is necessary to be aware of the fact that chemical state peak shifts occur which can be larger than 10 eV. For example the main Al peak in pure Al occurs at 1396 eV but in Al_2O_3 this peak occurs at 1378 eV, and this type of shift should be borne in mind when matching peaks. Clearly if peaks are small the elements are present in a relatively small amount and hence only the major peaks of these elements would be expected to be visible.

The Auger yield from a sample is defined by many factors some of which (fluorescent yield, ionization cross section) have been discussed in Chapter

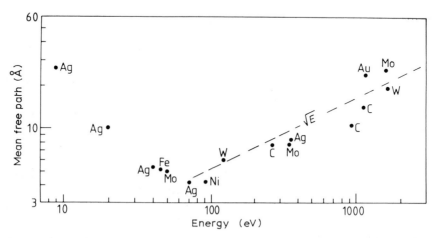

Fig. 6.9 Mean free path of electrons as a function of electron energy for various elements (after [26]).

2. In this section the various complications which arise in the analysis Auger data will be discussed. The first, and perhaps the most important parameter which influences the intensity of an Auger signal is the escape depth, i.e. the depth within a sample from which an Auger electron can escape from the sample and thus be detected. This escape depth is clearly directly related to the mean free path. The mean free path effectively defines the depth of the analysed volume and is clearly dependent on the energy of the Auger electron and the material making up the sample. In fact the energy dependence of the escape depth is a useful aid in determining the depth distribution of the elements near the surface. Typical experimental values for the mean free path of electrons are shown in Fig. 6.9. The range of the electrons increases roughly as $\sqrt{E}$ from about 75 eV.

Quantitative analysis is usually carried out by using the relative peak heights in the differentiated output from the various elements present, scaled by the appropriate intensity from a standard, which is usually the pure bulk element. On this basis the molar fraction X_A of element A in a sample would be given by

$$X_A = \frac{I_A/I_A^{st}}{\Sigma I_B/I_B^{st}} \tag{6.25}$$

where I_A and I_A^{st} are the appropriate respective intensities of a given Auger peak for element A from the sample and the standard. I_B and I_B^{st} are the appropriate respective intensities from all other elements in the sample and in standards. For a binary alloy of elements A and B the ratio X_A to X_B is thus given by

$$\frac{X_A}{X_B} = \frac{I_A/I_A^{st}}{I_B/I_B^{st}} \tag{6.26}$$

The values of I_A^{st}, I_B^{st}, etc. are commonly referred to as relative sensitivity factors and a summary taken from [19] is shown in Fig. 6.10 for the sensitivity factors for 5 and 10 kV incident electrons. These factors were obtained from pure elements or stoichiometric compounds.

These relative sensitivity factors I_A^{st}, etc. will not in general give the correct result becuase the Auger yield per atom from a pure sample will be different from that from an alloy because the matrix in an alloy will backscatter less or more efficiently than the pure metal. Thus if more electrons are backscattered more Auger electrons will be generated. For example in nickel–aluminium alloys the heavier nickel atom will backscatter more efficiently than aluminium atoms and hence increase the Al Auger yield relative to that in pure aluminium; the converse is true for the Ni Auger yield from nickel–aluminium alloys. The extent of the correction is clearly more important in materials which contain atoms of very different atomic weight. In addition

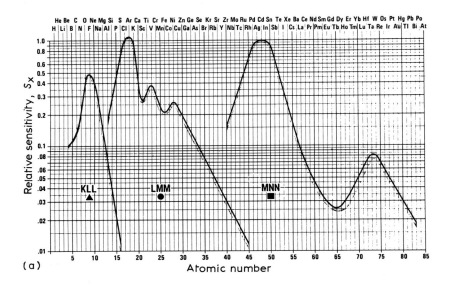

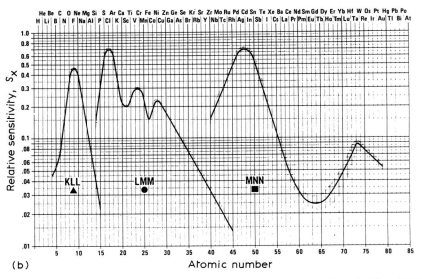

Fig. 6.10 Relative Auger sensitivity factors for the elements indicated (after [18]): (a) $E_p = 5$ keV; (b) $E_p = 10$ keV.

the form of the correction is determined by the type of sample under examination. Thus in the case of a homogenous binary alloy equation (6.26) becomes [20]

$$\frac{X_A}{X_B} = (F_{AB})\frac{I_A/I_A^{st}}{I_B/I_B^{st}} \quad (6.27)$$

where F_{AB} is given by

$$F_{AB}(X_A \to 0) = \frac{1 + r_A(E_A)}{1 + r_B(E_A)}\left(\frac{a_B}{a_A}\right)^{1.5} \quad (6.28)$$

$$F_{AB}(X_B \to 0) = \frac{1 + r_A(E_B)}{1 + r_B(E_B)}\left(\frac{a_B}{a_A}\right)^{1.5} \quad (6.29)$$

where $r_A(E_A)$ is the added contribution to the intensity of element A caused by backscattering and similarly $r_B(E_B)$. a_B and a_A are the atomic sizes of elements A and B. Because the numerical values for $F_{AB}(X_A \to 0)$ and $F_{AB}(X_B \to 0)$ are found to be very similar it is necessary to use only a single scaling factor for any pair of elements. The backscattering term can be evaluated using Monte Carlo calculations and an example of $r_A(E_A)$, the backscattering coefficient for each element plotted against the energy of the Auger electron, is shown in Fig. 6.11 for 5 kV electrons, incident at 30° to the specimen surface normal. Values of F_{AB} have been calulated for over four thousand element pairs and for a significant number the value of F_{AB} lies

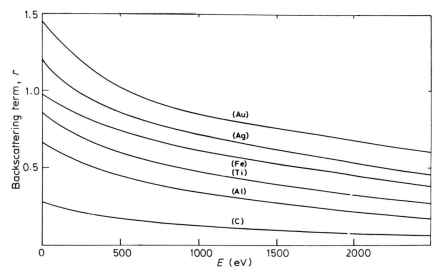

Fig. 6.11 Backscattering coefficient as a function of the energy of the Auger electrons for 5 kV electron incident at 30° to the specimen normal (after [19]).

in the range 2–3 and in dilute solutions the error caused by ignoring F_{AB} is virtually the value of F_{AB}.

Fig 6.11 is very useful as a guide to the likely value of F_{AB} for elements of known Z and as already indicated it is clear that the correction is most important when the two elements have very different atomic numbers.

The backscattering correction is different if the sample is a pure element B covered with a partial monolayer of element A. For low coverages the true fractional coverage ϕ_A can be shown [20] to be given by

$$\phi_A = Q_{AB} I_A / I_A^{st} \qquad (6.30)$$

where

$$Q_{AB} = \frac{\tau_A(E_A) \cos \theta}{Q_A} \left(\frac{1 + r_A(E_A)}{1 + r_B(E_B)} \right) \qquad (6.31)$$

$\tau_A(E_A)$ is the inelastic mean free path of electrons of energy E_A and $\cos \theta$ is the take-off angle. The mean free path can be approximated as $\tau_A = (0.41 \, a_A^{1.5} E_A^{1.5})$, where τ and a are in nanometres and E_A is in electronvolts. Substitution of typical values into the equation for Q_{AB} shows that this correction factor typically lies within the range 2–10 and hence this is an important. correction.

If ion beam milling is used to produce depth profiles then it is necessary to correct the analysis obtained at different milling times for the difference in sputtering rates of the elements in the sample. If the sputtering yields for elements A and B in a binary alloy are S_A and S_B then the steady state composition of the surface-depleted zone of an alloy after sputtering will be given by [20]

$$\bar{X}_A = X_A \left[X_A + \left(\frac{S_A}{N_A} \frac{N_B}{S_B} \right) X_B \right]^{-1} \qquad (6.32)$$

The true composition of the alloy X_A / X_B is then given by

$$\frac{X_A}{X_B} = F_{AB}^s \frac{I_A / I_A^{st}}{I_B / I_B^{st}} \qquad (6.33)$$

where

$$F_{AB}^s = \left(\frac{S_A}{N_A} \frac{N_B}{S_B} \right) F_{AB} \qquad (6.34)$$

In practice the correction is not as straightforward as implied by the above because the sputtering efficiencies of the pure elements can be very different from those same elements when they are present as oxides or other stable compounds. On this basis at least, ion beam depth profiles must be viewed with some caution and if accurate profiles are required then sputtering yields should be measured using standards in the form of oxides or whatever chemical state is appropriate. Additionally specimens trepanned with a hemispherical ball or taper sections may be used to obtain a depth profile.

Additional information similar to that in EELS is in fact available from Auger spectra because the precise energy of an Auger peak and its shape is a function of the chemical state of the atom giving rise to an Auger electron. This aspect of Auger spectroscopy is best discussed in conjunction with X-ray photoelectron spectroscopy (XPS) and it is not covered in this text.

6.6 SPATIAL RESOLUTION OF ANALYSIS

The spatial resolution of all X-ray and Auger analysis is influenced by high angle elastic scattering of electrons and may, under some conditions, be significantly degraded by X-ray fluorescence. On the other hand the spatial resolution of EELS is defined by the specimen thickness and the acceptance angle of the spectrometer. Thus in the case of EELS the spatial resolution δ from a specimen of thickness t is simply given by αt, where α is the collection angle. Electrons scattered through an angle such that they are not accepted by the spectrometer do not influence the spatial resolution. A typical foil thickness used for EELS would be 50 nm and even if α is as large as 50 mrad the spatial resolution would be 2.5 nm. In the case of EELS it is particularly easy to see what is meant by spatial resolution since only that part of the specimen within the cone defined by the collection angle will generate loss electrons which are collected by the spectrometer. Particles further apart than 2.5 nm will therefore be analysable separately and particles even closer than this could be analysed separately by reducing either the collection angle or sample thickness. Clearly the intensity of the signal will be decreased as α is reduced and the ultimate limit to the spatial resolution in EELS is set by signal to noise problems.

Although it is clear that high angle elastic scattering of electrons influences the spatial resolution of X-ray and Auger analysis it is by no means clear how the resolution can be defined. Thus the simplest single scattering models for X-ray analysis of thin foils ([21], [22] and [23]) are based on Rutherford scattering (see Chapter 2) and can be used to calculate the probe size on the bottom surface within which 90% (or some other defined percentage) of the electrons are contained. Thus the probe size b is given by [22]

$$b = \frac{Z(m/m_0)t\lambda^2}{\pi a_0 \sqrt{3}} \left(\frac{nt}{\pi p}\right)^{1/2} \tag{6.35}$$

where Z is the atomic number; m and m_0 are the dynamic and rest masses of the electron; t is the foil thickness; n is the number of atoms per unit volume; a_0 is the Bohr radius (~ 53 pm); λ is the wavelength of the electrons and $p = 0.1$ if b is to contain 0.9 of the scattered electrons, i.e. if b is to be the

Table 6.3 Values of the 90% probe size calculated using equation (6.35) for 100 kV electrons.

Element	Foil thickness (nm)		
	50	100	300
Al	3.0	8.0	42.0
Cu	7.5	21.5	112.0
Au	17.0	—	—

90% probe size. Thus

$$b \propto \frac{Zt^{3/2}}{E} \tag{6.36}$$

and the probe size increases linearly with atomic number and with (thickness)$^{3/2}$ and is inversely proportional to accelerating voltage E. Typical values of b for 100 kV electrons for several elements are shown in Table 6.3 for a point incident beam. No values are shown for gold foils thicker than 100 nm because multiple scattering would be expected to invalidate the single scattering model [21]. Whilst equation (6.35) is in reasonable agreement with experimental observations [24], [25] and also reveals the important factors which influence high angle elastic scattering, it does not easily enable an assessment of whether two precipitates could be separately analysed. Thus it is by no means obvious if the 90% probe size is b that two particles a distance b apart could be analysed totally independently. Some fraction of the 10% of electrons which are scattered through large angles will pass through the particle outside the 90% probe and would generate X-rays, perhaps of sufficient intensity to be detected (see Chapter 2 for detection limits). The probability of degradation of the resolution is of course greater if the two particles are near the electron exit surface of the specimen. The spatial resolution is in fact definable only if complete three-dimensional information of the sample is available together with appropriate Monte Carlo calculations. This is neither possible in most cases nor justified and a more useful approximate definition of the spatial resolution δ is given by [23]

$$\delta^2 = d_i^2 + b^2 \tag{6.37}$$

where d_i is the 90% incident probe size and b the 90% probe size calculated from equation (6.35). Only in very rare instances will more accurate assessments of the spatial resolution be required. In STEM a typical probe size would be 4 nm for a LaB$_6$ filament and equation (6.37) gives the spatial resolution as 22 nm for a 100 nm thick copper sample at 100 kV. For realistic samples the initial STEM probe size is then of secondary importance.

In the case of bulk samples in Auger and X-ray analysis the spatial resolution can again be accurately calculated only if Monte Carlo calculations are carried out. The resolution is worse the higher the voltage and the lighter the atomic weight (i.e. the opposite to the case for X-ray analysis of thin foils) because the electrons will penetrate further and generate X-rays and Auger electrons at larger distances from the initial probe position. In the case of Auger spectroscopy it is backscattered electrons, which generate Auger electrons sufficiently near the surface for them to escape, that limit the spatial resolution. The only way to improve the spatial resolution is to reduce the voltage of the incident electrons but for efficient X-ray and Auger production the overvoltage should be at least two (see Fig. 2.14), and this sets a lower limit to the useable voltage. The spatial resolution of a 30 kV microprobe is typically about 2 μm whereas the lateral resolution of Auger spectroscopy is about 50 nm and is better both because a lower voltage is used and because the generation of Auger electrons below the surface is irrelevant.

REFERENCES

1. Cliff, G. and Lorimer, G.W. (1975) *J. Microscopy*, **103**, 203.
2. Duncumb, P. (1968) *J. de Microscope*, 7, 581.
3. Philibert, J. and Tixier, R. (1968) *Br. J. Appl. Phys.*, **1**, 685.
4. Goldstein, J.I., Costley, J.L., Lorimer, G.W. and Reed, S.J.B. (1977) *Scanning Electron Microsc.*, **1**, 315.
5. Bracewell, B.L. and Veigle, W.J. (1971) in *Developments in Applied Spectroscopy*, Vol. 9, (ed. A. Perkins), Plenum Press, New York.
6. Nockolds, C., Nasir, M.J., Cliff, G. and Lorimer, G.W. (1980) in *Electron Microscopy and Analysis 1979:Inst. Phys. Conf. Ser. No. 52*, (ed. T. Mulvey), Adam Hilger, Bristol, p. 417.
7. Nicholls, A.W. and Jones, I.P. (1982) *J. Microscopy*, **127**, 119.
8. Lorimer, G.W., Al-Salman, S.A. and Cliff, G. (1977) in *Developments in Electron Microscopy and Analysis 1977: Inst. Phys. Conf. Ser. No. 36*, (ed. D.L. Misell), Adam Hilger, Bristol, p. 369.
9. Hall, C.R. (1966) *Proc. R. Soc.* A, **295**, 140.
10. Castaing, J. and Descamps, J. (1955) *J. Phys. Rad.*, **16**, 304.
11. Reed, S.J.B. (1975) *Electron Probe Microanalysis*, Cambridge University Press, Cambridge.
12. Brown, L.M. (1981) *J. Phys. F: Metal Phys.*, **11**, 1.
13. Zaluzec, N.J. (1981) *Library of Electron Energy Loss Spectra*, Argonne National Laboratory Publication, Chicago, Illinois.
14. Egerton, R.F. and Whelan, M.J. (1974) *Phil. Mag.*, **30**, 739.
15. Spence, J.C.H. (1979) *Ultramicroscopy*, **4**, 9.
16. Egerton, R.F. (1979) *Ultramicroscopy*, **4**, 169.
17. Leapman, R.D., Rez, P. and Mayers, D.F. (1980) *J. Chem. Phys.*, **72**, 1232.
18. Joy, D.C. and Maher, D.M. (1982) *J. Microscopy*, **124**, 37.
19. Davis, L.E., McDonald, C., Palmberg, P.W., Riach, G.E. and Weber, R.E. (1976)

Handbook of Auger Electron Spectroscopy, 2nd edn, Physical Electronics Industries Inc., Minnesota, USA.
20. Seah, M.P. (1981) *Analysis*, **9**, 171.
21. Goldstein, J.I., Costley, J.L., Lorimer, G.W. and Reed, S.J.B. (1971) in *Scanning Electron Microscopy*, Vol. 1, (ed. O. Johari) Chicago Press, Chicago, p. 315.
22. Jones, I.P. and Loretto, M.H. (1981) *J. Microscopy*, **124**, 3.
23. Reed, S.J.B. (1982) *Ultramicroscopy*, **7**, 405.
24. Hutchings, R., Loretto, M.H., Jones, I.P. and Smallman, R.E. (1979) *Ultramicroscopy*, **3**, 401.
25. Stephenson, T., Loretto, M.H. and Jones, I.P. (1981) *Quantitative Microanalysis with High Spatial Resolution*, Metals Society, London.
26. Chang, C.C. (1974) *Characterisation of Solid Surfaces*, Plenum Press, New York.

Appendix A

THE RECIPROCAL LATTICE

When the translations of a primitive space lattice are denoted by **a**, **b** and **c**, the vector **p** to any lattice point is given $\mathbf{p} = u\mathbf{a} + v\mathbf{b} + w\mathbf{c}$. The definition of the reciprocal lattice is that the translations **a***, **b*** and **c***, which define the reciprocal lattice fulfil the following relationships:

$$\mathbf{a}^* \cdot \mathbf{a} = \mathbf{b}^* \cdot \mathbf{b} = \mathbf{c}^* \cdot \mathbf{c} = K^2 = 1 \tag{A.1}$$

$$\mathbf{a}^* \cdot \mathbf{b} = \mathbf{b}^* \cdot \mathbf{c} = \mathbf{c}^* \cdot \mathbf{a} = \mathbf{a} \cdot \mathbf{b}^* = \mathbf{b} \cdot \mathbf{c}^* = \mathbf{c} \cdot \mathbf{a}^* = 0 \tag{A.2}$$

It can then be easily shown that:
(1) $|\mathbf{c}^*| = 1/c$-spacing of primitive lattice and similarly for **b*** and **a***.
(2) $\mathbf{a}^* = (\mathbf{b} \wedge \mathbf{c})/\mathbf{a} \cdot (\mathbf{b} \wedge \mathbf{c})$ $\mathbf{b}^* = (\mathbf{c} \wedge \mathbf{a})/\mathbf{b} \cdot (\mathbf{c} \wedge \mathbf{a})$ $\mathbf{c}^* = (\mathbf{a} \wedge \mathbf{b})/\mathbf{c} \cdot (\mathbf{a} \wedge \mathbf{b})$

These last three relations are often used as a definition of the reciprocal lattice.

Two properties of the reciprocal lattice are particularly important: (a) The vector **g*** defined by $\mathbf{g}^* = h\mathbf{a}^* + k\mathbf{b}^* + l\mathbf{c}^*$ (where h, k and l are integers to the point hkl in the reciprocal lattice) is normal to the plane of Miller indices (hkl) in the primary lattice. (b) The magnitude $|\mathbf{g}^*|$ of this vector is the reciprocal of the spacing of (hkl) in the primary lattice.

A.1 ALLOWED REFLECTIONS

The kinematical structure factor for reflections is given by

$$F_{hkl} = \sum_i f_i \exp[-2\pi i(hu_i + kv_i + lw_i)] \tag{A.3}$$

where u_i, v_i and w_i are the coordinates of the atoms and hkl the Miller indices of the reflection **g**.

If there is only one atom at 0, 0, 0 in the unit cell then the structure factor will be independent of hkl since for all values of h, k and l we have $F_{hkl} = f$. Thus if b.c.c. and f.c.c. crystals are referred to their primitive cells then reflections from the simple metals such as niobium and copper, which have only one atom at each lattice point, will all have intensities given by $f(\theta)^2$. Since the three basic **g** vectors derived below, for the b.c.c. structure for

Table A.1 Necessary conditions for allowed reflections in terms of the type of unit cells in crystals.

Unit cell	Possible reflections	Forbidden reflections
Primitive	All values of h, k and l	none
Body centred	$(h+k+l)$ even	$(h+k+l)$ odd
Face centred	h, k, l all odd or even	h, k and l mixed
Base centred	h and k both odd or even	h and k mixed

example, are $[110]^*$, $[011]^*$ and $[101]^*$ it is clear that the allowed reflections are those derived by summing or subtracting these vectors, i.e. reflections such that $(h+k+l)$ is always even. A similar conclusion is reached if the cubic structure cell, which contains atoms at $0, 0, 0$ and $\frac{1}{2}, \frac{1}{2}, \frac{1}{2}$, is used since

$$F_{hkl} = f\left\{\exp(-2\pi i 0) + \exp\left[-2\pi i\left(\frac{h}{2} + \frac{k}{2} + \frac{l}{2}\right)\right]\right\}$$

$$= \begin{cases} f(1+1) = 2f & \text{if } (h+k+l) \text{ is even} \\ f(1-1) = 0 & \text{if } (h+k+l) \text{ is odd.} \end{cases}$$

The allowed reflections in the various possible unit cells are shown in Table A.1

If the number of atoms in a unit cell is large there is the possibility that some of the allowed reflections will have zero intensity, e.g. the 200 reflection in silicon. Nevertheless all the allowed reflections must conform with the above classifications.

It is of course an important part of structure determination to recognize which reflections are absent (see Chapter 4). The absent reflections in metals such as zirconium (i.e. those for which $(h+2k)$ is a multiple of 3 and l is odd) show that the structure is h.c.p..

A.2 RECIPROCAL LATTICE FOR F.C.C. AND B.C.C. CRYSTALS

Referring the primitive translations to cubic axis for a primitive f.c.c. cell we have

$$\mathbf{a} = \frac{a}{2}[110] = \frac{a}{2}[\mathbf{i}+\mathbf{j}] \quad \mathbf{b} = \frac{a}{2}[011] = \frac{a}{2}[\mathbf{j}+\mathbf{k}] \quad \mathbf{c} = \frac{a}{2}[101] = \frac{a}{2}[\mathbf{i}+\mathbf{k}]$$

where $\mathbf{i}$, $\mathbf{j}$ and $\mathbf{k}$ are unit vectors along the cubic axes. Since

$$\mathbf{a}^* = \frac{\mathbf{b} \wedge \mathbf{c}}{\mathbf{a} \cdot (\mathbf{b} \wedge \mathbf{c})}$$

then

$$\mathbf{a}^* = \frac{\frac{1}{2}a[\mathbf{j}+\mathbf{k}] \wedge \frac{1}{2}a[\mathbf{i}+\mathbf{k}]}{\frac{1}{2}a[\mathbf{i}+\mathbf{j}] \cdot \frac{1}{2}a[\mathbf{j}+\mathbf{k}] \wedge \frac{1}{2}a[\mathbf{i}+\mathbf{k}]}$$

and

$$\frac{\frac{1}{4}a^2[\mathbf{i}+\mathbf{j}-\mathbf{k}]}{\frac{1}{8}a^3[\mathbf{i}+\mathbf{j}] \cdot [\mathbf{i}+\mathbf{j}-\mathbf{k}]} = \frac{1}{a}[\mathbf{i}+\mathbf{j}-\mathbf{k}] = \frac{1}{a}[11\bar{1}]$$

Correspondingly,

$$\mathbf{b}^* = \frac{1}{a}[\bar{1}11] \quad \text{and} \quad \mathbf{c}^* = \frac{1}{a}[1\bar{1}1]$$

with a b.c.c. structure and referring translations to cubic axes then

$$\mathbf{a} = \frac{a}{2}[11\bar{1}] \quad \mathbf{b} = \frac{a}{2}[\bar{1}11] \quad \mathbf{c} = \frac{a}{2}[1\bar{1}1]$$

so that

$$\mathbf{a}^* = \frac{\frac{1}{2}a[-\mathbf{i}+\mathbf{j}+\mathbf{k}] \wedge \frac{1}{2}a[\mathbf{i}-\mathbf{j}+\mathbf{k}]}{\frac{1}{2}a[\mathbf{i}+\mathbf{j}-\mathbf{k}] \cdot \frac{1}{2}a[-\mathbf{i}+\mathbf{j}+\mathbf{k}] \wedge \frac{1}{2}a[\mathbf{i}-\mathbf{j}+\mathbf{k}]}$$

which reduces to

$$\frac{\frac{1}{4}a^2[\mathbf{i}+\mathbf{j}]}{\frac{1}{8}a^3[\mathbf{i}+\mathbf{j}-\mathbf{k}] \cdot [\mathbf{i}+\mathbf{j}]} = \frac{1}{a}[110]$$

Similarly,

$$\mathbf{b}^* = \frac{1}{a}[011] \quad \text{and} \quad \mathbf{c}^* = \frac{1}{a}[101]$$

Note that the basic vectors in the reciprocal lattice for these non-primitive cells are twice those of the crystal of the same system. Thus, the basic translations are $a/2 \langle 111 \rangle$ for a b.c.c. crystal and $1/a \langle 111 \rangle$, rather than $1/2a \langle 111 \rangle$, for the reciprocal lattice for an f.c.c. crystal.

A.3 DOUBLE DIFFRACTION

Because electrons are strongly scattered it is possible that rescattering of a diffracted beam can give rise to a strong diffracted beam where structure factor considerations suggest the beam should be of zero intensity.

The conditions under which a forbidden diffraction spot may appear are most easily seen using the Ewald sphere construction (see Chapter 4). Thus, if the reciprocal lattice point corresponding to $\mathbf{g}_1$ for which the structure factor is large, lies on the Ewald sphere then a strong diffracted beam will be produced in the direction $\mathbf{A}_1$, as shown on Fig. A.1. Similarly if $\mathbf{g}_2$ also

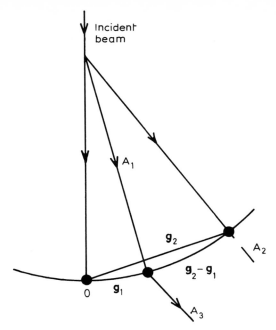

Fig. A.1 Ewald sphere construction showing the conditions which must be fulfilled for double diffraction to occur. See text.

lies on the sphere a diffracted beam would be expected in the direction A_2. However, if the structure factor is zero for $\mathbf{g}_2$ the intensity of A_2 would be zero but rediffraction of the beam A_1 by planes perpendicular to $\mathbf{g}_2 - \mathbf{g}_1$ will give rise to a beam A_3 which is parallel to and therefore indistinguishable from A_2.

Diffraction maxima due to double diffraction can be distinguished by rotating the crystal about the direction which contains the spot in question. The intensity of this spot will be unchanged unless it is due to double diffraction when it will disappear when $\mathbf{g}_1$ is no longer excited.

A.4 SHAPES OF DIFFRACTION MAXIMA

Significant diffracted intensity is observed from thin samples even when the Bragg condition is not precisely satisfied ([1], [2]). It can be shown that only for an infinite crystal will the diffraction maxima be points and that as the crystal dimensions get smaller so the diffraction maxima get larger. For a parallelepiped crystal the diffracted amplitude is given by

$$\phi_g = \frac{F_g}{V_c} \int_0^A \int_0^B \int_0^C \exp[-2\pi i(ux + vy + wz)]\,dx\,dy\,dz \qquad \text{(A.4)}$$

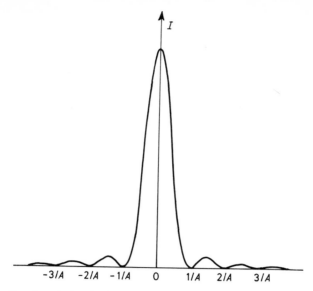

Fig. A.2 Predicted variation in intensity of a diffracted beam from a crystal of edge length A as a function of deviation from the Bragg condition where the deviation is measured in units of $1/A$. See text.

$$= \frac{F_g}{V_c} \frac{\sin(\pi A u)}{\pi u} \frac{\sin(\pi B v)}{\pi v} \frac{\sin(\pi C w)}{\pi w} \quad \text{(A.5)}$$

where A, B and C are the edge lengths of the parallelepiped along x, y and z, and u, v, w are the values of $\mathbf{s}$ (the deviation from the Bragg condition) along x, y and z. Fig. A.2 shows how the intensity (obtained by multiplying ϕ_g by ϕ_g^*, its complex conjugate), varies along u for v and $w = 0$. The central maximum has a width at half maximum height of $1/A$ and successive minima occur at intervals of $1/A$. The intensity of the successive maxima decreases very rapidly, as shown in Fig. A.2 but because electron diffraction patterns have such a large dynamic range it is possible to detect maxima out to at least the fiftieth maximum for typical thickness samples.

For very thin crystals, such as small precipitates, the spikes in reciprocal space give rise to visible intensity for all beam directions. The dimension of the central diffraction maximum along a given direction in reciprocal space is given by $1/d$ where d is the parallel dimension of the specimen. Thus a thin plate precipitate gives rise to a long spike in reciprocal space normal to the plate and a needle precipitate gives rise to a disc of intensity.

REFERENCES

1. Hirsch, P.B., Howie, A., Nicholson, R.B., Pashley, D.W. and Whelan, M.J. (1965) *Electron Microscopy of Thin Crystals*, Butterworths, Sevenoaks.
2. James, R.W. (1958) *Optical Principles of the Diffraction of X-rays*, Bell, London.

Appendix B

INTERPLANAR DISTANCES AND ANGLES IN CRYSTALS. CELL VOLUMES. DIFFRACTION GROUP SYMMETRIES

B.1 THE SEVEN SYSTEMS

The axial lengths and angles and the symmetries exhibited in the seven crystal systems are shown in Table B.1.

Table B.1 The axial relationships and symmetries of the seven crystal systems.

Crystal	Axial length and angles	Minimum symmetry elements
Cubic	Three equal axes at right angles $a = b = c, \alpha = \beta = \gamma = 90°$	Four, threefold rotation axes
Hexagonal	Two coplanar axes at 120°, third axis at right angles $a = b \neq c, \alpha = \beta = 90°, \gamma = 120°$	One, sixfold rotation (or rotation-inversion) axis
Trigonal (or rhombohedral)	Three equal axes equally inclined $a = b = c, \alpha = \beta = \gamma \neq 90°$	One, threefold rotation (or rotation-inversion) axis
Tetragonal	Three axes at right angles $a = b \neq c, \alpha = \beta = \gamma = 90°$	One, fourfold rotation (or rotation-inversion) axis
Orthorhombic	Three orthogonal unequal axes $a \neq b \neq c, \alpha = \beta = \gamma = 90°$	Three, perpendicular twofold (or rotation-inversion) axis
Monoclinic	Three unequal axes one pair not orthogonal $a \neq b \neq c, \alpha = \gamma = 90° \neq \beta$	One, twofold rotation (or rotation-inversion) axis

Table B.1 (*Contd.*)

Triclinic	Three unequal axes none at right angles $a \neq b \neq c, \alpha \neq \beta \neq \gamma \neq 90°$	None

B.2 INTERPLANAR SPACING

The value of d, the distance between adjacent planes in the set (hkl), may be found from the following equations.

Cubic: $\quad \dfrac{1}{d^2} = \dfrac{h^2 + k^2 + l^2}{a^2}$

Tetragonal: $\quad \dfrac{1}{d^2} = \dfrac{h^2 + k^2}{a^2} + \dfrac{l^2}{c^2}$

Hexagonal: $\quad \dfrac{1}{d^2} = \dfrac{4}{3}\left(\dfrac{h^2 + hk + k^2}{a^2}\right) + \dfrac{l^2}{c^2}$ (Miller indices)

Rhombohedral: $\quad \dfrac{1}{d^2} = \dfrac{(h^2 + k^2 + l^2)\sin^2 \alpha + 2(hk + kl + hl)(\cos^2 \alpha - \cos \alpha)}{a^2(1 - 3\cos^2 \alpha + 2\cos^3 \alpha)}$

Orthorhombic: $\quad \dfrac{1}{d^2} = \dfrac{h^2}{a^2} + \dfrac{k^2}{b^2} + \dfrac{l^2}{c^2}$

Monoclinic: $\quad \dfrac{1}{d^2} = \dfrac{1}{\sin^2 \beta}\left(\dfrac{h^2}{a^2} + \dfrac{k^2 \sin^2 \beta}{b^2} + \dfrac{l^2}{c^2} - \dfrac{2hl \cos \beta}{ac}\right)$

Triclinic: $\quad \dfrac{1}{d^2} = \dfrac{1}{V^2}(S_{11}h^2 + S_{22}k^2 + S_{33}l^2 + 2S_{12}hk + 2S_{23}kl + 2S_{13}hl)$

In the equation for triclinic crystals,

V = volume of unit cell (see below)

$S_{11} = b^2 c^2 \sin^2 \alpha$

$S_{22} = a^2 c^2 \sin^2 \beta$

$S_{33} = a^2 b^2 \sin^2 \gamma$

$S_{12} = abc^2 (\cos \alpha \cos \beta - \cos \gamma)$

$S_{23} = a^2 bc (\cos \beta \cos \gamma - \cos \alpha)$

$S_{13} = ab^2 c (\cos \gamma \cos \alpha - \cos \beta)$

B.3 INTERPLANAR ANGLES

The angle ϕ between the plane $(h_1 k_1 l_1)$, of spacing d_1, and the plane $(h_2 k_2 l_2)$, of spacing d_2, may be found from the following equations (V is the volume of the unit cell).

Cubic:
$$\cos \phi = \frac{h_1 h_2 + k_1 k_2 + l_1 l_2}{[(h_1^2 + k_1^2 + l_1^2)(h_2^2 + k_2^2 + l_2^2)]^{1/2}}$$

Tetragonal:
$$\cos \phi = \left(\frac{h_1 h_2 + k_1 k_2}{a^2} + \frac{l_1 l_2}{c^2} \right)$$
$$\times \left[\left(\frac{h_1^2 + k_1^2}{a^2} + \frac{l_1^2}{c^2} \right) \left(\frac{h_2^2 + k_2^2}{a^2} + \frac{l_2^2}{c^2} \right) \right]^{-1/2}$$

Hexagonal:
$$\cos \phi = \left(h_1 h_2 + k_1 k_2 + \tfrac{1}{2}(h_1 k_2 + h_2 k_1) + \frac{3a^2}{4c^2} l_1 l_2 \right)$$
$$\times \left[\left(h_1^2 + k_1^2 + h_1 k_1 + \frac{3a^2}{4c^2} l_1^2 \right) \right.$$
$$\left. \times \left(h_2^2 + k_2^2 + h_2 k_2 + \frac{3a^2}{4c^2} l_2^2 \right) \right]^{-1/2}$$

Rhombohedral:
$$\cos \phi = \frac{a^4 d_1 d_2}{V^2} [\sin^2 \alpha (h_1 h_2 + k_1 k_2 + l_1 l_2)$$
$$+ (\cos^2 \alpha - \cos \alpha)$$
$$\times (k_1 l_2 + k_2 l_1 + l_1 h_2 + l_2 h_1 + h_1 k_2 + h_2 k_1)]$$

Orthorhombic:
$$\cos \phi = \left(\frac{h_1 h_2}{a^2} + \frac{k_1 k_2}{b^2} + \frac{l_1 l_2}{c^2} \right)$$
$$\times \left[\left(\frac{h_1^2}{a^2} + \frac{k_1^2}{b^2} + \frac{l_1^2}{c^2} \right) \left(\frac{h_2^2}{a^2} + \frac{k_2^2}{b^2} + \frac{l_2^2}{c^2} \right) \right]^{-1/2}$$

Monoclinic:
$$\cos \phi = \frac{d_1 d_2}{\sin^2 \beta} \left[\frac{h_1 h_2}{a^2} + \frac{k_1 k_2 \sin^2 \beta}{b^2} + \frac{l_1 l_2}{c^2} \right.$$
$$\left. - \frac{(l_1 h_2 + l_2 h_1) \cos \beta}{ac} \right]$$

Triclinic:
$$\cos \phi = \frac{d_1 d_2}{V^2} [S_{11} h_1 h_2 + S_{22} k_1 k_2 + S_{33} l_1 l_2$$
$$+ S_{23}(k_1 l_2 + k_2 l_1) + S_{13}(l_1 h_2 + l_2 h_1)$$
$$+ S_{12}(h_1 k_2 + h_2 k_1)]$$

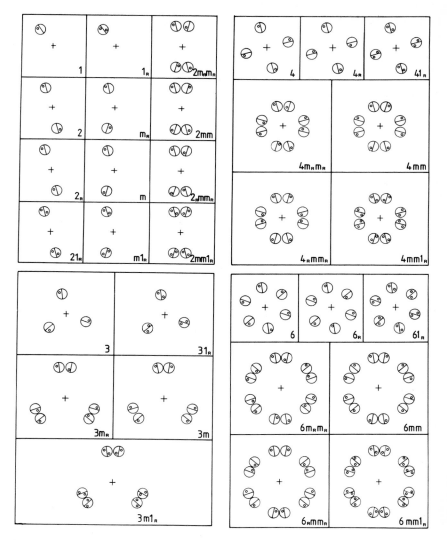

Fig. B.1 Figures illustrating the symmetry properties of the 31 diffraction groups (see text).

B.4 CELL VOLUMES

The following equations give the volume V of the unit cell.

Cubic: $\quad V = a^3$

Tetragonal: $\quad V = a^2 c$

Hexagonal: $\quad V = \dfrac{a^2 c \sqrt{3}}{2} = 0.866 a^2 c$

Rhombohedral: $\quad V = a^3 (1 - 3\cos^2 \alpha + 2\cos^3 \alpha)^{1/2}$

Orthorhombic: $\quad V = abc$

Monoclinic: $\quad V = abc \sin \beta$

Triclinic: $\quad V = abc(1 - \cos^2 \alpha - \cos^2 \beta - \cos^2 \gamma + 2\cos\alpha \cos\beta \cos\gamma)^{1/2}$

B.5 DIFFRACTION GROUP SYMMETRIES

Schematic drawings illustrating the symmetries of the 31 diffraction groups, (taken from [1]) are shown in Fig. B.1. The number refers to the rotation symmetry, for example all the groups beginning with 3 exhibit some form of threefold rotation symmetry. The letter m denotes the presence of a vertical mirror plane, the group $3m1_R$ for example has a mirror which bisects the individual pairs of dark field maxima in the figure. The subscript R is used to indicate that the operation which has a subscript associated with it requires a rotation of π about the individual point in the pattern, after the subscripted operation has been performed, in order to restore the initial symmetry. For example if the diffraction group 6 is compared with the diffraction group 6_R it is clear that the original pattern is restored in diffraction group 6 simply by rotating the whole pattern in increments of $2\pi/6$ but the individual symbols must be rotated by π after each whole pattern rotation of $2\pi/6$ for the 6_R group. The symbol 1_R means that the original symmetry is retained if each individual symbol is rotated by π so that the diffraction group 6_R therefore does not possess this property whereas the 61_R does.

REFERENCES

1. Buxton, B.F., Eades, J.A., Steeds, J.W. and Rackham, G.M. (1976) *Phil. Trans. R. Soc.* A, **281**, 171.

Appendix C

KIKUCHI MAPS, STANDARD DIFFRACTION PATTERNS AND EXTINCTION DISTANCES

The schematic Kikuchi maps and diffraction patterns which are shown in this appendix (Figs. C.1–C.15) have been indexed so that the indices of poles correspond to the upward drawn directions. The indices of the Kikuchi

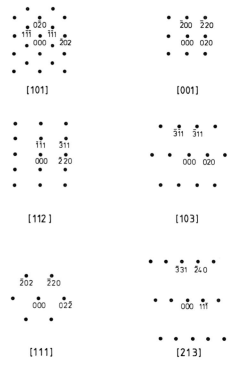

Fig. C.1 Schematic diffraction patterns for six selected electron beam directions for a f.c.c. crystal. The low order reflections are indexed and the upward drawn direction, i.e. the electron beam direction **B** is indicated under each diffraction pattern.

190 ELECTRON BEAM ANALYSIS OF MATERIALS

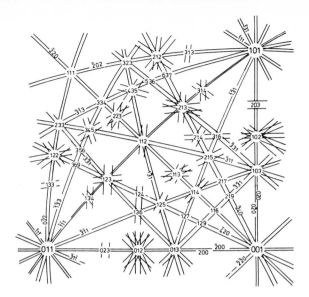

Fig. C.2 Schematic Kikuchi map for a f.c.c. crystal extending over two standard triangles.

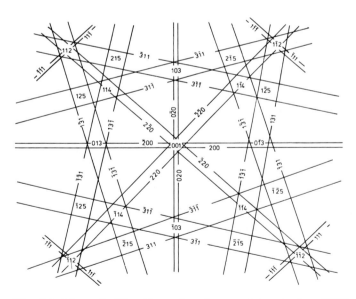

Fig. C.3 Schematic Kikuchi map for a f.c.c. crystal centred on [001].

APPENDIX C   191

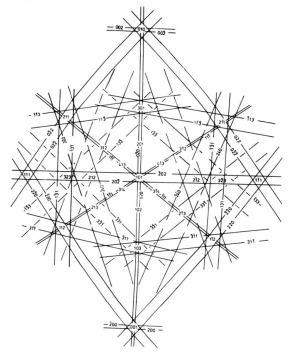

Fig. C.4 Schematic Kikuchi map for a f.c.c. crystal centred on [101].

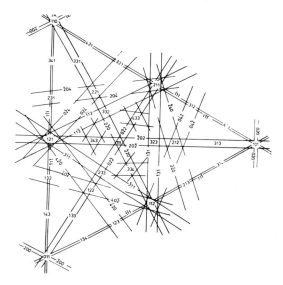

Fig. C.5 Schematic Kikuchi map for a f.c.c. crystal centred on [111].

192 ELECTRON BEAM ANALYSIS OF MATERIALS

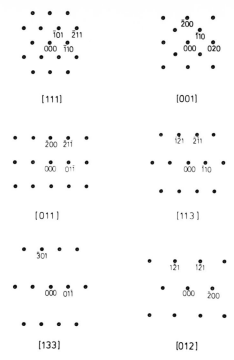

Fig. C.6 Schematic diffraction patterns for six selected electron beam directions for a b.c.c. crystal. The low order reflections are indexed and the upward drawn direction, i.e. the electron beam direction **B** is indicated under each diffraction pattern.

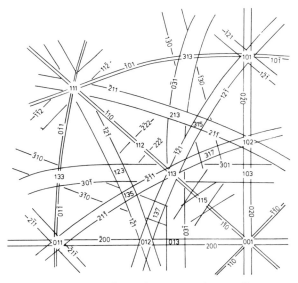

Fig. C.7 Schematic Kikuchi map for a b.c.c. crystal extending over two standard triangles.

APPENDIX C 193

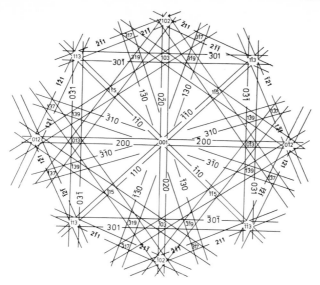

Fig. C.8 Schematic Kikuchi map for a b.c.c. crystal centred on [001].

Fig. C.9 Schematic Kikuchi map for a b.c.c. crystal centred on [011].

194 ELECTRON BEAM ANALYSIS OF MATERIALS

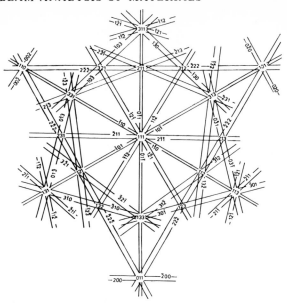

Fig. C.10 Schematic Kikuchi map for a b.c.c. crystal centred on [111].

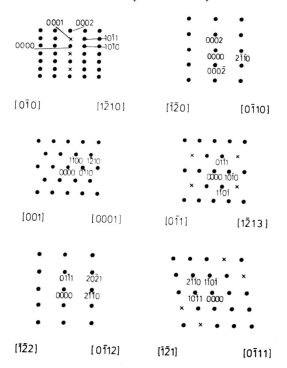

Fig. C.11 Schematic electron diffraction patterns for six selected electron beam directions for a h.c.p. crystal. The low order reflections are indicated and the Miller indices and Miller–Bravais indices of the upward drawn direction, i.e. the electron beam direction **B** is indicated below each pattern. The crosses indicate reflections which can occur due to double reflection.

APPENDIX C

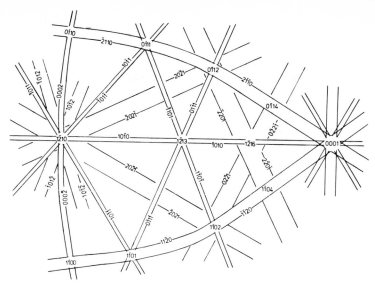

Fig. C.12 Schematic Kikuchi map for a h.c.p. crystal extending over two standard triangles.

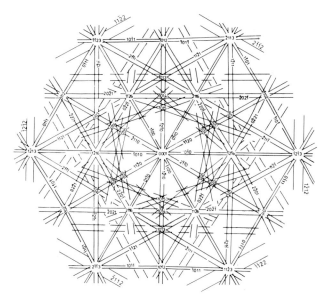

Fig. C.13 Schematic Kikuchi map for a h.c.p. crystal centred on [0001].

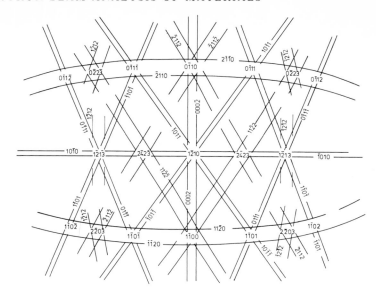

Fig. C.14 Schematic Kikuchi map for a h.c.p. crystal centred on $[1\bar{2}10]$.

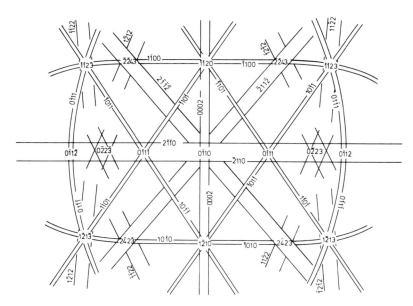

Fig. C.15 Schematic Kikuchi map for a h.c.p. crystal centred on $[0\bar{1}10]$.

APPENDIX C   197

lines appear on the lines themselves with the sign of **g** marked to correspond to the indices of the diffraction maximum through which the line passes when the Bragg condition is satisfied. For example, considering the 020 reflection in the f.c.c. map for $\mathbf{B} = [001]$ the 020 Kikuchi line would pass through the 020 maximum when the 020 reflection is at the Bragg condition so that the line marked $0\bar{2}0$ would then pass through the origin of the reciprocal lattice, the direct beam. If the crystal is tilted appropriately the situation will be reversed and the $0\bar{2}0$ reflection will be at the Bragg condition and the 020 Kikuchi line will pass through the direct beam.

It should be emphasized that the maps are not drawn to scale and should be regarded as topologically correct only. If detailed work is to be carried out on any particular crystal system it is a very valuable exercise to make up a composite experimental Kikuchi map.

The usefulness of reflections in diffraction contrast is effectively measured by the extinction distance for the reflection (see Chapter 2) and a few selected values are given in Table C.1 for 100 kV electrons. To obtain the extinction distances for other voltages these values should be scaled by the ratio of electron velocities.

Table C.1 Typical extinction distances for 100 kV electrons (in nm).

Reflection	Al	Cu	Ni	Au	Si	Ge
111	56	24	24	16	60	43
200	67	28	28	18	—	—
220	106	42	41	25	76	45
311	130	51	50	29	135	76
222	138	54	53	31	—	—
400	167	65	65	36	127	66

Reflection	Fe	Nb	Mo
110	27	26	23
200	39	37	32
211	50	46	41
310	71	62	58

Reflection	Mg	Co	Zn	Zr	Cd
0002	81	25	26	32	24
$1\bar{1}01$	100	31	35	38	32
$11\bar{2}0$	141	43	50	49	44
$1\bar{1}00$	151	47	55	59	52
$11\bar{2}2$	171	52	58	59	68
$2\bar{2}01$	202	62	70	69	61
$1\bar{1}02$	231	70	76	84	68

Appendix D
STEREOMICROSCOPY AND TRACE ANALYSIS

D.1 STEREOMICROSCOPY

Stereomicroscopy is used in both SEM and TEM to obtain qualitative and quantitative three-dimensional information from specimens. Stereo pairs are taken simply by tilting the specimen between micrographs. In the case of SEM stereopairs this is a very straightforward procedure since all that is required is to operate the appropriate graduated tilt control after first taking a micrograph and taking a second micrograph at this different perspective. In the case of TEM using diffraction contrast imaging (Chapter 5) it is essential to change only **B** and to maintain **g** and s_g identical for the two micrographs. Clearly this is best done using Kikuchi maps and tilting along the appropriate Kikuchi line. The tilt angle can be calculated from the angle between the two beam directions.

Qualitative three-dimensional information can be obtained by viewing the stereomicrographs in a stereoviewer with the tilt axis towards the viewer. Fracture surfaces in SEM and the defect arrangement within a thin foil in TEM and in particular HVEM are far more easily interpreted when viewed in stereo. Quantitative three-dimensional information can be obtained by using the height measuring attachments available on stereoviewers. The apparent height difference measured on the viewer δ is related to real height difference H through the tilt angle θ and the magnification M:

$$H = \frac{\delta}{2M \sin \theta}$$

D.2 TRACE ANALYSIS

When appropriate crystallographic information is available it is straightforward to relate line directions of defects or directions in surfaces to the orientation of the crystal axes. For unambiguous analysis it is clear that information is required in more than one projection so that true directions can be extracted from projected directions. Thus in the case of the projected

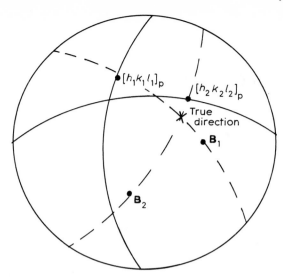

Fig. D.1 Schematic stereogram illustrating the technique of trace analysis to determine the true direction of a line which when viewed in $\mathbf{B}_1$ and $\mathbf{B}_2$ projects along $[h_1k_1l_1]_p$ and $[h_2k_2l_2]_p$ respectively.

line direction $[h_1k_1l_1]_p$ of a dislocation viewed in a direction $\mathbf{B}_1$ all that can be said is that the true direction lies in the plane defined by $[h_1k_1l_1]_p$ and $\mathbf{B}_1$. Micrographs taken in any other beam direction $\mathbf{B}_2$ result in a second projected direction $[h_2k_2l_2]_p$ and hence a plane defined by $[h_2k_2l_2]_p$ and by $\mathbf{B}_2$. The zone axis for these two planes defines the true direction of the dislocation. This is illustrated in Fig. D.1.

Appendix E
TABLES OF X-RAY AND EELS ENERGIES

Tables of X-ray and EELS energies are shown on the following pages.

APPENDIX E 201

Table E.1 Energies (in keV) of X-rays from the elements tabulated as a function of increasing energy.

Energy	Element	Atomic number	Principal emission line	Less intense minor line			Lesser intense minor line			Even less intense minor line		
				Energy	Line	Intensity	Energy	Line	Intensity	Energy	Line	Intensity
0.851	Ni	28	L_α									
0.883	Ce	58	M_α									
0.930	Cu	29	L_α									
0.972	Ba	56	M_γ									
1.012	Zn	30	L_α									
1.041	Na	11	K_α									
1.081	Sm	62	M_α									
1.098	Ga	31	L_α									
1.185	Gd	64	M_α									
1.188	Ge	32	L_α									
1.253	Mg	12	K_α									
1.282	As	33	L_α									
1.379	Se	34	L_α									
1.480	Br	35	L_α									
1.486	Al	13	K_α									
1.521	Yb	70	M_α									
1.586	Kr	36	L_α									
1.644	Hf	72	M_α									
1.694	Rb	37	L_α									
1.709	Ta	73	M_α									
1.739	Si	14	K_α									
1.774	W	74	M_α									
1.806	Sr	38	L_α									
1.842	Re	75	M_α									
1.914	Os	76	M_α									
1.922	Y	39	L_α									
1.977	Ir	77	M_α									

Table E.1 (Contd.)

Energy	Element	Atomic number	Principal emission line	Less intense minor line			Lesser intense minor line			Even less intense minor line		
				Energy	Line	Intensity	Energy	Line	Intensity	Energy	Line	Intensity
2.013	P	15	K_{α_1}	2.028	K_{α_2}	(10)	2.137	K_β	(6)			
2.042	Zr	40	L_{α_1}	2.124	L_{β_1}	(45)						
2.048	Pt	78	M_α	2.127	M_β	(50)						
2.121	Au	79	M_α	2.204	M_β	(50)						
2.166	Nb	41	L_{α_1}	2.257	L_{β_1}	(45)	2.163	L_{α_2}	(10)			
2.195	Hg	80	M_α	2.282	M_β	(50)						
2.267	Tl	81	M_α	2.362	M_β	(55)						
2.293	Mo	42	L_{α_1}	2.394	L_{β_1}	(45)						
2.307	S	16	K_{α_1}	2.322	K_{α_2}	(50)	2.465	K_β	(10)			
2.342	Pb	82	M_α	2.442	M_β	(60)						
2.419	Bi	83	M_α	2.524	M_β	(60)						
2.558	Ru	44	L_{α_1}	2.683	L_{β_1}	(45)						
2.621	Cl	17	K_{α_1}	2.631	K_{α_2}	(10)	2.815	K_β	(8)			
2.696	Rh	45	L_{α_1}	2.834	L_{β_1}	(40)	3.001	L_{β_2}	(25)			
2.838	Pd	46	L_{α_1}	2.990	L_{β_1}	(40)	3.171	L_{β_2}	(25)			
2.957	Ar	18	K_{α_1}	3.190	K_β	(15)						
2.984	Ag	47	L_{α_1}	3.150	L_{β_1}	(40)	3.347	L_{β_2}	(25)			
2.991	Th	90	M_α	3.145	M_β	(60)	3.369	M_γ	(5)			
3.077	Pa	91	M_α	3.239	M_β	(60)	3.465	M_γ	(5)			
3.133	Cd	48	L_{α_1}	3.316	L_{β_1}	(42)	3.528	L_{β_2}	(25)			
3.165	U	92	M_α	3.336	M_β	(60)	3.563	M_γ	(5)			
3.286	In	49	L_{α_1}	3.487	L_{β_1}	(75)	3.713	L_{β_2}	(17)			
3.312	K	19	K_α	3.589	K_β	(15)						
3.443	Sn	50	L_{α_1}	3.662	L_{β_1}	(75)	3.904	L_{β_2}	(17)			
3.604	Sb	51	L_{α_1}	3.843	L_{β_1}	(75)	4.100	L_{β_2}	(17)			
3.690	Ca	20	K_α	4.012	K_β	(15)						
3.769	Te	52	L_α	4.029	L_{β_1}	(75)	4.301	L_{β_2}	(17)			
3.937	I	53	L_α	4.220	L_{β_1}	(75)	4.507	L_{β_2}	(17)			
4.088	Sc	21	K_α	4.460	K_{β_1}	(20)						
4.109	Xe	54	L_α	4.42	L_{β_1}	(50)	4.72	L_{β_2}	(20)			

APPENDIX E 203

		Z							
4.508	Ti	22	K_α	4.931	K_{β_1}	(20)			
4.650	La	57	L_α	5.041	L_{β_1}	(50)	5.383	L_{β_2}	(20)
4.839	Ce	58	L_α	5.261	L_{β_1}	(50)	5.612	L_{β_2}	(20)
4.949	V	23	K_α	5.426	K_{β_1}	(20)			
5.411	Cr	24	K_α	5.946	K_{β_1}	(18)			
5.635	Sm	62	L_α	6.204	L_{β_1}	(50)	6.586	L_{β_2}	(20)
5.894	Mn	25	K_α	6.489	K_{β_1}	(20)			
6.398	Fe	26	K_α	7.057	K_{β_1}	(20)			
6.924	Co	27	K_α	7.648	K_{β_1}	(20)			
7.471	Ni	28	K_α	8.263	K_{β_1}	(20)			
7.898	Hf	72	L_α	9.021	L_{β_1}	(50)	9.346	L_{β_2}	(20)	10.514	L_{γ_1}	(10)
8.040	Cu	29	K_α	8.904	K_{β_1}	(20)			
8.145	Ta	73	L_α	9.342	L_{β_1}	(50)	9.650	L_{β_2}	(20)	10.893	L_{γ_1}	(10)
8.396	W	74	L_α	9.671	L_{β_1}	(50)	9.960	L_{β_2}	(20)	11.284	L_{γ_1}	(10)
8.630	Zn	30	K_α	9.570	K_{β_1}	(20)			
8.651	Re	75	L_α	10.008	L_{β_1}	(50)	10.274	L_{β_2}	(20)	11.683	L_{γ_1}	(10)
8.910	Os	76	L_α	10.354	L_{β_1}	(50)	10.597	L_{β_2}	(20)	12.093	L_{γ_1}	(10)
9.174	Ir	77	L_α	10.706	L_{β_1}	(50)	10.919	L_{β_2}	(20)	12.510	L_{γ_1}	(10)
9.241	Ga	31	K_α	10.262	K_{β_1}	(21)			
9.441	Pt	78	L_α	11.069	L_{β_1}	(50)	11.249	L_{β_2}	(20)	12.940	L_{γ_1}	(10)
9.712	Au	79	L_α	11.440	L_{β_1}	(50)	11.583	L_{β_2}	(20)	13.379	L_{γ_1}	(10)
9.874	Ge	32	K_α	10.979	K_{β_1}	(21)			
9.987	Hg	80	L_α	11.821	L_{β_1}	(50)	11.922	L_{β_2}	(20)	13.828	L_{γ_1}	(10)
10.267	Tl	81	L_α	12.211	L_{β_1}	(50)	12.270	L_{β_2}	(20)	14.289	L_{γ_1}	(10)
10.542	As	33	K_α	11.722	K_β	(22)			
10.550	Pb	82	L_α	12.612	L_{β_1}	(50)	12.621	L_{β_2}	(20)	14.762	L_{γ_1}	(10)
10.837	Bi	83	L_α	13.021	L_{β_1}	(50)	12.978	L_{β_2}	(20)	15.245	L_{γ_1}	(10)
11.207	Se	34	K_α	12.492	K_β	(24)			
11.907	Br	35	K_α	13.286	K_β	(24)			
12.631	Kr	36	K_α	14.107	K_β	(24)			
12.967	Th	90	L_α	16.199	L_{β_1}	(50)	15.621	L_{β_2}	(20)	18.979	L_{γ_1}	(10)
13.373	Rb	37	K_{α_1}	14.956	K_β	(24)			
13.612	U	92	L_α	17.217	L_{β_1}	(50)	16.425	L_{β_2}	(20)	20.164	L_{γ_1}	(10)
14.140	Sr	38	K_{α_1}	15.830	K_β	(24)			
14.931	Y	39	K_α	16.734	K_β	(25)			
15.744	Zr	40	K_α	17.663	K_β	(27)			
16.581	Nb	41	K_α	18.700	K_β	(12)			

Table E.1 (Contd.)

Energy*	Individual energies	Element	Atomic number	Principal emission line	Less intense minor line			Less intense minor line		
					Energy	Line	Intensity	Energy	Line	Intensity
17.441	(17.476,17.371)	Mo	42	$K_{\alpha_{1,2}}$	19.599	K_β	(25)			
19.233	(19.276,19.147)	Ru	44	$K_{\alpha_{1,2}}$	21.646	K_β	(24)			
20.165	(20.213,20.070)	Rh	45	$K_{\alpha_{1,2}}$	22.712	K_β	(24)			
21.121	(21.174,21.017)	Pd	46	$K_{\alpha_{1,2}}$	23.807	K_β	(26)			
22.101	(22.159,21.987)	Ag	47	$K_{\alpha_{1,2}}$	24.921	K_β	(26)			
23.106	(23.170,22.980)	Cd	48	$K_{\alpha_{1,2}}$	26.080	$K_{\beta_{1,3}}$	(27)	26.639	K_{β_2}	(5)
24.136	(24.206,23.998)	In	49	$K_{\alpha_{1,2}}$	27.252	$K_{\beta_{1,3}}$	(27)	27.856	K_{β_2}	(5)
25.191	(25.267,25.040)	Sn	50	$K_{\alpha_{1,2}}$	28.467	$K_{\beta_{1,3}}$	(28)	29.104	K_{β_2}	(5)
26.271	(26.355,26.105)	Sb	51	$K_{\alpha_{1,2}}$	29.705	$K_{\beta_{1,3}}$	(29)	30.388	K_{β_2}	(5)
28.508	(28.607,28.312)	I	53	$K_{\alpha_{1,2}}$	32.271	$K_{\beta_{1,3}}$	(29)	33.036	K_{β_2}	(6)
29.666	(29.774,29.453)	Xe	54	$K_{\alpha_{1,2}}$	33.598	$K_{\beta_{1,3}}$	(29)	34.408	K_{β_2}	(6)
30.851	(30.968,30.620)	Cs	55	$K_{\alpha_{1,2}}$	34.962	$K_{\beta_{1,3}}$	(30)	35.815	K_{β_2}	(6)
32.062	(32.188,31.812)	Ba	56	$K_{\alpha_{1,2}}$	36.317	$K_{\beta_{1,3}}$	(28)	37.251	K_{β_2}	(7)
33.299	(33.436,33.028)	La	57	$K_{\alpha_{1,2}}$	37.771	$K_{\beta_{1,3}}$	(30)	38.723	K_{β_2}	(7)
34.566	(34.714,34.273)	Ce	58	$K_{\alpha_{1,2}}$	39.232	$K_{\beta_{1,3}}$	(32)			
39.911	(40.111,39.516)	Sm	62	$K_{\alpha_{1,2}}$						

* Weighted average energy of K_{α_1} and K_{α_2}. Individual energies of unresolved peaks listed in next column.

Table E.2 Energies (in keV) of characteristic X-rays of the elements.

Element	Atomic number	K_α	K_β	(*)	L_α	L_{β_1}	(*)	L_{β_2}	(*)	L_γ	(*)	M
Na	11	1.041										
Mg	12	1.253										
Al	13	1.486										
Si	14	1.739										
P	15	2.013	2.137	(6)								
S	16	2.307	2.465	(10)								
Cl	17	2.621	2.815	(8)								
Ar	18	2.957	3.190	(15)								
K	19	3.312	3.589	(15)								
Ca	20	3.690	4.012	(15)	0.341							
Sc	21	4.088	4.460	(20)	0.395							
Ti	22	4.508	4.931	(20)	0.452							
V	23	4.949	5.426	(20)	0.511							
Cr	24	5.411	5.946	(18)	0.573							
Mn	25	5.894	6.489	(20)	0.637							
Fe	26	6.398	7.057	(20)	0.705							
Co	27	6.924	7.648	(20)	0.733							
Ni	28	7.471	8.263	(20)	0.851							
Cu	29	8.040	8.804	(20)	0.930							
Zn	30	8.630	9.570	(20)	1.012							
Ga	31	9.241	10.262	(21)	1.098							
Ge	32	9.874	10.979	(21)	1.188							
As	33	10.542	11.722	(22)	1.282							
Se	34	11.207	12.492	(24)	1.379							
Br	35	11.907	12.286	(24)	1.480							
Kr	36	12.631	14.107	(24)	1.586							
Rb	37	13.373	14.956	(24)	1.694							
Sr	38	14.140	15.830	(24)	1.806							
Y	39	14.931	16.734	(25)	1.922							
Zr	40	15.744	17.633	(27)	2.042	2.124	(45)					

Table E.2 (Contd.)

Element	Atomic number	K_α	K_β	(*)	L_α	L_{β_1}	(*)	L_{β_2}	(*)	L_γ	(*)	M
Nb	41	16.581	18.700	(12)	2.166	2.257	(45)					0.355
Mo	42	17.441	19.599	(25)	2.293	2.394	(45)					0.331
Tc	43	18.325	20.608	(24)	2.424	2.536	(45)					
Ru	44	19.233	21.646	(24)	2.558	2.683	(45)					0.461
Rh	45	20.165	22.712	(24)	2.696	2.834	(40)					0.496
Pd	46	21.121	23.807	(26)	2.838	2.990	(40)					0.532
Ag	47	22.101	24.921	(26)	2.984	3.150	(40)	3.001	(25)			0.568
Cd	48	23.106	26.167	(27)	3.133	3.316	(42)	3.171	(25)			0.606
In	49	24.136	27.346	(27)	3.286	3.487	(75)	3.347	(25)			
Sn	50	25.191	28.564	(28)	3.443	3.662	(75)	3.528	(25)			0.691
Sb	51	26.271	29.805	(28)	3.604	3.843	(75)	3.713	(17)			0.733
Te	52	27.377	31.097	(29)	3.769	4.029	(75)	3.904	(17)			0.778
I	53	28.508	32.402	(29)	3.937	4.220	(75)	4.100	(17)			
Xe	54	29.666	33.737	(30)	4.109	4.42	(50)	4.301	(17)			
Cs	55	30.851	35.104	(30)	4.286	4.619	(50)	4.507	(17)			
Ba	56	32.062	36.504	(31)	4.465	4.829	(50)	4.72	(20)			0.972
La	57	33.299	37.951	(31)	4.650	5.041	(50)	4.935	(20)			0.833
Ce	58	34.566	39.232	(32)	4.839	5.261	(50)	5.193	(20)			0.883
Pr	59	35.860			5.033	5.488	(50)	5.383	(20)			0.929
Nd	60	37.182			5.229	5.721	(50)	5.612	(20)			0.978
Pm	61	38.532			5.432	5.960	(50)	5.849	(20)			
Sm	62	39.911			5.635	6.204	(50)	6.088	(20)			1.081
Eu	63				5.845	6.455	(50)	6.338	(20)			1.131
Gd	64				6.056	6.712	(50)	6.586	(20)			1.185
Tb	65				6.272	6.977	(50)	6.842	(20)			1.240
Dy	66				6.494	7.246	(50)	7.102	(20)			1.293

APPENDIX E

Element	Z								
Ho	67	6.719	7.524	(50)	7.910	(20)			1.347
Er	68	6.947	7.809	(50)	8.188	(20)			1.405
Tm	69	7.179	8.100	(50)	8.467	(20)	9.424	(5)	1.462
Yb	70	7.414	8.400	(50)	8.757	(20)	9.778	(5)	1.521
Lu	71	7.654	8.708	(50)	9.038	(20)	10.142	(6)	1.581
Hf	72	7.898	9.021	(50)	9.346	(20)	10.514	(10)	1.644
Ta	73	8.145	9.342	(50)	9.650	(20)	10.893	(10)	1.709
W	74	8.396	9.671	(50)	9.960	(20)	11.284	(10)	1.774
Re	75	8.651	10.008	(50)	10.274	(20)	11.683	(10)	1.842
Os	76	8.910	10.354	(50)	10.597	(20)	12.093	(10)	1.914
Ir	77	9.174	10.706	(50)	10.919	(20)	12.510	(10)	1.977
Pt	78	9.441	11.069	(50)	11.249	(20)	12.940	(10)	2.074
Au	79	9.712	11.440	(50)	11.583	(20)	13.379	(10)	2.148
Hg	80	9.987	11.821	(50)	11.922	(20)	13.828	(10)	2.224
Tl	81	10.267	12.211	(50)	12.270	(20)	14.289	(10)	2.301
Pb	82	10.550	12.612	(50)	12.621	(20)	14.762	(10)	2.380
Bi	83	10.837	13.021	(50)	12.978	(20)	15.245	(10)	2.458
Po	84	11.129	13.445	(50)	13.338	(20)	15.741	(10)	
At	85	11.425	13.574	(50)	14.065	(10)	16.249	(10)	
Rn	86	11.725	14.313	(50)	14.509	(10)	16.768	(10)	
Fr	87	12.029	14.768	(50)	14.448	(20)	17.300	(10)	
Ra	88	12.338	15.233	(50)	14.839	(20)	17.845	(10)	
Ac	89	12.650	15.710	(50)	15.929	(10)	18.405	(10)	
Th	90	12.967	16.199	(50)	15.621	(20)	18.979	(10)	3.058
Pa	91	13.288	16.699	(50)	16.022	(20)	19.565	(10)	3.148
U	92	13.612	17.217	(50)	16.425	(20)	20.164	(10)	3.239
Np	93	13.942	17.747	(50)	16.837	(20)	20.781	(10)	

(*) Approximate intensity relative to principal line of series.

Table E.3 Energies (in eV) corresponding to edges in EELS data.

	K	L_3	L_2		L_3	L_2	M_5	M_4
Li	55	—	—	As	1323	1359	—	—
Be	111	—	—	Se	1435	1476	57	57
B	188	—	—	Br	1550	1596	69	70
C	284	—	—	Kr	1675	1727	89	89
N	402	—	—	Rb	1804	1864	110	112
O	532	—	—	Sr	1940	2007	133	135
F	685	—	—	Y	2080	2156	157	160
Ne	867	—	—	Zr	2222	2307	150	152
Na	1072	—	—	Nb	—	—	205	207
Mg	1305	51	51	Mo	—	—	227	230
Al	1560	73	73	Tc	—	—	253	256
Si	1839	99	99	Ru	—	—	279	284
P	2146	132	132	Rh	—	—	307	312
S	—	165	165	Pd	—	—	335	340
Cl	—	200	202	Ag	—	—	367	373
Ar	—	245	247	Cd	—	—	404	411
K	—	294	296	In	—	—	443	451
Ca	—	346	350	Sn	—	—	485	493
Sc	—	402	407	Sb	—	—	528	537
Ti	—	456	462	Te	—	—	572	583
V	—	513	521	I	—	—	619	631
Cr	—	575	584	Xe	—	—	672	—
Mn	—	640	651	Cs	—	—	726	740
Fe	—	708	721	Ba	—	—	781	796
Co	—	779	794	La	—	—	832	846
Ni	—	855	872	Ce	—	—	883	901
Cu	—	931	951	Sm			1080	1106
Zn	—	1020	1043					
Ga	—	1115	1142					
Ge	—	1217	1248					

INDEX

Aberration
 chromatic, 4, 10
 focussing error, 43
 spherical, 2, 10
Allowed reflections, 179
Antiphase boundary, 140
Aperture
 condenser aperture in convergent beam, 90
 diffraction limit, 11
 objective aperture, 40
Auger electrons, generation of, 35
Auger electron spectroscopy, 16, 53
 interpretation of spectra in, 169

Backscattering of electrons, 30
 images, 149
 patterns, 112
Black/white contrast, 132
Bragg's law, 68
Bremsstrahlung, see X-rays
Burgers vector, FS/RH definition, 124
 determination of, 120
 loop analysis, 124

Channelling patterns, 109
Cluster formation, influence on diffraction patterns, 104
Contrast transfer function, 146
Convergence, 13
 in SEM, 46
Convergent beam diffraction, 52, 79
 foil thickness determination, 80
 point group determination, 90
 space group determination, 95
 unit cell determination, 95
Coster–Kronig transition, 34
Coupling in energy loss, diffraction, 61, 62
 image, 61, 62
Cross section for ionization, 28
 partial, 167
Crystal planes, angles between, 186
 spacing of, 70, 185
Crystal system, determination of, 66, 76

Debye–Waller factor, 23

Deflection systems, 17
Detector
 energy dispersive, 56, 59
 scintillation, 46
 wavelength dispersive, 56, 57
Deviation parameter, 75, 114, 128
Diffraction contrast, in TEM, 114
Diffraction from aperture, 10
Diffraction group symmetry, 88, 90, 188
Diffraction patterns, 40, 65
 analysis of, 62, 65
 errors in SAD, 45
 from twinned crystals, 108
 rotation with respect to image, 42
Diffuse scattering, 103
Dislocation images, 119
Dispersion of spectrometer, 16
Double diffraction, 181
Dynamical theory, 117

Elastic scattering, of electrons, 19
 high angle, 23
 low angle, 22
 Rutherford scattering, 20, 24
Electron beam induced current signal, 37
Electron energy loss spectroscopy, 15, 163
Electron lenses, 11
Electron microprobe analysis, 56
Electron sources, 1
Electron spectrometer
 magnetic electron, 61
 Auger electron, 63
Ewald sphere, 68
Extinction distance, 22
 effective, 23, 119

Field depth, in SEM, 46
Fluorescent yield, 28, 32

High order Laue zones, 82
High resolution electron microscopy, 43, 144
High voltage electron microscopy, 43
HT stability, 4

Inelastic scattering, of electrons, 24

INDEX

plasmon, 25
single electron, 25
thermal, 24

Kikuchi lines, 72
Kikuchi maps, 189
Kinematic theory, 114

Lattice, Bravais, 22
Laue zones, indexing of, 82
Lenses
  in Auger spectroscopy, 16
  auxiliary, 12
  condenser, 12
  electrostatic, 7
  focal length, 10
  image forming, 11
  magnetic, 7
  probe forming, 11, 14
  rotation, 10
Long range order, 66, 101

Magnetic domains, in TEM, 142
  in SEM, 151
Magnetic prism, 15
Magnetic samples, 108
  imaging of, 142
Miller indices, 70
Miller Bravais indices, 72
Moiré patterns, 131
Monte Carlo calculations, 31

Omega transformation, 107
Optics, electron, 7

Partial dislocation analysis, 141
Planar defect, contrast from, 134
Precipitation and diffuse scattering, 103
Premartensitic phenomena, 105
Pretransformation and diffuse scattering, 103, 105
Probe size, in SEM, 47

Reciprocal lattice, 68, 179
Resolution
  of Auger spectroscopy, 55, 175
  of back scattered images, 50
  of electron energy loss spectroscopy, 175
  of energy loss spectrometer, 16
  of secondary images, 48
  of X-ray analysis, 50, 53, 175
  of X-ray spectrometer, 59, 61
Rocking curves, 115

Scanning electron microscopy, 14, 44
  interpretation of images, 147

Scanning transmission electron microscopy, 11, 13, 50, 143
Short range order, 102
Source, electron, 3
  brightness of, 2, 7
  coherence, 4
  energy spread, 4
  field emission, 6
  LaB_6, 2, 6
  size, 3, 4, 7
  stability
  thermionic, 4
Spatial resolution of analysis, 175
Spectrometer, magnetic, 61
  Auger, 63
Spinodal decomposition, 104
Stacking fault contrast analysis, 134
  in f.c.c. crystals, 137
  in h.c.p. crystals, 139
Stereomicroscopy, 198
Stopping power, 31
Strain contrast, from precipitates, 138
Structure factor, 22, 181

Thickness fringes, 116
Trace analysis, 198
ω-Transformation, 107
Transmission electron microscopy, 39
  analysis of images in, 119
  diffraction contrast in, 114
  influence of electron optical conditions, 143
Tunnelling, electron, 2
Twinning, 108

Weak beam microscopy, 128
Wehnelt cap, 4, 6

Void contrast, 130

X-rays
  absorption and fluorescence of, 157
  bremsstrahlung, 34
  characteristic, 27
  cross section for ionization, 28
  fluorescent yield, 28
  from bulk samples, 162
  generation of, 26
  influence of diffraction condition on production, 162
  interpretation of X-ray data from thin foils, 153
X-ray spectrometer
  wavelength dispersive, 57
  energy dispersive, 59